From
THE BIG BANG
to
HEAVEN

A new concept of the universe

Hassan E. Sajadi

Copyright © 2012 Hassan E. Sajadi
All rights reserved.
ISBN: 146797627X
ISBN 13: 9781467976275

Dedication:

To my children, Ali and Reza

To my dear grandchildren, Maël and Laia, and those who may come…

To my spouse, Shahnaz

Table of Contents

Preface vii

Introduction   Modern Cosmology, a brief history ix

Chapter I   **Major related theories** 1
  Special relativity 1
  General relativity 4
  Quantum theory 8

Chapter II   **The origin of the world** 12
  Planck era, a lawless land 15
  Out of nothing 17

Chapter III   **Big Bang: The Classic Model** 21
  Evidences for the big bang 23
  Unsolved problems with standard model 26

Chapter IV   **Big Bang: A New Model,
  our Conception** 33
  Escape velocity 33
  Thesis postulate 37

	The Proposed Model of the Universe 38
	The primary bang..38
	The redshift..44
	Age of the universe 48
	Phase of continuous expansion 51
Chapter V	Major cosmological events 59
	Soon after the big bang 59
	All in the middle...61
	Birth and death of stars and galaxies.............62
	Gravitational collapse and black holes 63
	Black holes in our concept of universal escape speed 67
Chapter VI	Time 71
	Newton's fixed time......................................73
	Einstein's flexi time 74
	A clock in the sky... 75
	Our conceptual vision of time:.....................77
	Past, present, and future 77
	The final end...84

Preface

I am not a writer, nor a physicist; this is why I must start by apologizing for the shortcomings that the reader may find both in language and substance while reading this book. My reason for writing it is to attempt to introduce and share a *new concept* in what is now considered as the "hot big bang (BB) model"; the actually accepted version. This model, as we all know, presents many fundamental problems and inconsistencies for which no definite answer is yet proposed. Chief among them are the problems of the age of the universe, its fate, and its uniformity (known as the *horizon* problem). In our conceptual model, as we shall see, those problems simply do not arise.

The concept is proposed in all simplicity, without hard physical and mathematical formula. In fact a conscious attempt has been made throughout the book to avoid the use of complex physical and mathematical jargon and formulae. The aim has been to make the many complex notions and physical theories involved in the subject of cosmology accessible to the average curious reader.

The second aim in writing this book is to reassemble dispreads works on such a fundamental and fascinating subject, and to present it in a concise, short manner without technical scientific complexity or deep and heavy philosophical notions. It intends to be an overall view of the different physical notions and theories involved in the description and comprehension of the science of cosmology. It is mainly aimed at ordinary readers who do not have time to explore fundamental theories in physics in any depth.

This book consists of six major parts. The first two parts introduce the three fundamental theories in physics, which involve the description of cosmology and the origin of the world—the theories of relativity, gravitation, and quantum physics. The third part describes the currently accepted classic BB model, with a presentation of the principal arguments for and against it. The forth part presents the body of our concept: a new model of thinking regarding the big bang. The concept eliminates all the inconsistencies of the classic model. The proposed view is further developed in chapter five, which deals mainly with the chronology of major cosmological events and the description of the life cycle of stars and galaxies.

The sixth part, a separate chapter, is allocated to the topic of time, because time is intimately and inexorably intermixed in all physical theories describing cosmology. This chapter completes our conceptual views on the evolution of the universe and also on the fascinating subject of time itself.

The choice of the title of the book will become apparent at the end.

Introduction

Modern Cosmology, a brief history

Cosmology, the science of the world, is the scientific study of the large-scale properties of the universe. Using the scientific method, the discipline's goals are to understand the origin, evolution, and the ultimate outcome of the entire universe. Cosmology, maybe more than other scientific fields, has a long history into the past, often mixed with philosophy and mythology. Kepler, Galileo, Newton, and others have contributed to its more recent development; but more than anyone else, it is Einstein, who contributed most enormously in the understanding of modern cosmology. Cosmology's development has mainly consisted of the formation of theories and hypotheses based on current scientific data, and on making specific predictions that could be later tested by observation. These theories are constantly modified, abandoned, or revised in the light of new observations. Modern cosmology starts with the publication of Einstein's theory of Relativity.

Here is a chronological list of the major events of modern cosmological development:

1905—Albert Einstein publishes the special theory of relativity, positing that space and time are not separate entities but a continuum. Three consequences:

- Constancy of speed of light regardless of movement or position

- Flexible time and space. Four dimensional space-time continuums

- Mass-energy relationship $E=MC^2$

1915—Albert Einstein publishes the general theory of relativity. Its outcomes:

- Gravitation: the force of gravity working at distance and instantaneously! This is inconsistent with his theory of special relativity.

- To overcome this inconsistency, Einstein argued that gravity is not a force, but a distortion of space…a curvature of space.

- The principle of the equivalence of gravity and acceleration.

- A universe in a finite spherical form. What, then, surrounds this finite spherical universe? Einstein used his spherical ellipsoidal geometry of general relativity to propose curved space. What stops a finite spherical universe from gravitationally collapsing? Einstein proposed his cosmological antigravity constant, the lambda force.

1922—Friedmann, the Russian mathematician and meteorologist, realized that Einstein's equations could describe an expanding universe. Einstein was reluctant, believing in a static universe.

1929—The American astronomer Hubble established that some nebulae (fuzzy patches of light on the night sky) were indeed distant galaxies comparable in size to our own Milky Way. Hubble discovered the redshift with distance. If Doppler phenomenon caused this redshift, then it meant that stars and galaxies were moving apart. This was interpreted as evidence that the universe is expanding. Einstein, swayed by this argument, changed his mind—thus his comment about his "bigger blunder", referring to the cosmological constant.

1950—The British astronomer Fred Hoyle dismissively coined the phrase "big bang", and the name was widely adopted. The theory argued that the universe had been born at one moment, about some thousand million years ago and that the galaxies were still travelling away from us after that initial burst. All the matter, indeed the universe itself, was created at just one instant.

1965—Penzias and Wilson discovered a cosmic microwave background radiation, which had been predicted by Alpher and Hermann back in 1949. This was interpreted as the faint afterglow of the intense radiation of a hot big bang.

Since the **1970s** almost all cosmologists have come to accept the hot big bang model. More than anyone else, it was Einstein who contributed to the development of modern cosmology and the BB theory.

Chapter I

Major Related Theories

The theory of "Relativity", the theory of "Gravitation", and the theory of "Quantum" form together the fundamental backgrounds for understanding the new cosmology and the hot BB model. We will therefore review very briefly these three theories before describing the classic BB model.

Special Relativity

Isaac Newton founded his classic mechanics on the view that *space* is something distinct from the existing bodies, and it would always be there in the absence of all existing material. He thought that *time* was something that passes uniformly and inexorably no matter what happens in the world. He spoke of *absolute space* and *absolute time.* Later, by adding *weight* and *force,* and through their interactions, *motion*, he constructed a monumental system of mechanics explaining the physical world. It was a solid theoretical structure, which persisted for over two hundred years. But already during Newton's own time, and soon after him, contrary views were expressed, denying space as an independent entity. It was argued that an empty space had no meaning; that it was a conceptual impossibility.

In fact, the central principle of *inertia*, known since Galileo and developed by Newton, indicates that if a mass is not subject to any force acting upon it, the mass is either in a state of total rest or in uniform motion along a straight line. In fact a *uniform motion in straight line* is totally equivalent to a *rest situation.*

In order to illustrate, let us imagine the case of a boat moving with a uniform speed in a straight line alongside a quay or some other platform. You are inside the boat and have no access to the outside view. It would seem to you as if the boat was completely at rest; the same feeling that one has inside a cruising plane. Inside the boat all physical happenings are taking place as if the boat was in total arrest: drops of condensation from the ceiling are falling on a vertical line for example, and if there are some butterflies in the cabin, their flight seems totally normal. They all fly more or less at the same speed, no matter their direction.

What is the view of someone standing on the platform—someone who can see the inside? Let us suppose that the boat is moving at two meters per second (2m/s), and the butterflies at 1m/s. He would notice that first the drops from the ceiling are not falling on a vertical line to the platform but have a rather curved line of fall; and second, that the speed of the butterflies is different. Those flying in the direction of the boat's motion have a speed of 3m/s, and those in the opposing direction fly at 1m/s; a logical addition of the boat speed and butterflies proper speed.

By analogy, scientists in the beginning of nineteenth century have taken Earth as a boat moving in space; the Newtonian absolute space. According to the above logic, the speed of a beam of light propagating in the direction of the Earth's motion should be greater than that of a beam going the opposite way, as the speed of the Earth motion should be logically added or subtracted.

Or, through many incontestable and repeated experiments, it was proved that this was not the case, and that the speed of the light beam was exactly the same no matter its direction. The speed of the moving ship—

the Earth—did not change the light's speed by an iota. What happened to the logical speed-adding rule? The result of experiments that were repeated so many times was not contestable; neither was the physical law of speed addition. The deadlock was total.

It was Einstein who solved the dilemma by dismissing the notion of absolute time. The universality of time had never been questioned before. He suggested that the *seconds* inside and outside the boat must be different. Time was not the same for all observers, and in all places. For anyone, his own time.

Motion influences time, decreasing its rate. Time slows down at high speeds. The reason why this is not perceptible is that, in everyday life, the speeds are so low that the changes are extremely small. When we approach relativistic speeds, the effect becomes substantial. At the speed of light, it is dramatic, as the time dilation is infinite. Time stops—it no longer flows! Thus, the speed of light becomes a limit: nothing can go faster. We have all heard of the famous travelling twin returning from a long cosmic trip in a space ship travelling at a relatively high speed, who finds she has aged less than her sister remaining on earth.

With the fall of the theory of absolute time, out went also the notion of absolute space.

The theory of relativity does away with the notion of space and time as separate and absolute entities and introduces the notion of a space-time continuum. It fixes the speed of light as the upper speed limit.

This theory, a massive intellectual earthquake, revolutionized physics of the twentieth century and introduced fundamental changes in the perception of reality. But soon Einstein found it incomplete, which is why he called it "special relativity." He set out to find the complete theory of "general relativity". That proved to be a much harder exercise.

General Relativity

Gravity, a force of attraction known by Galileo and others before him, was described scientifically by Newton in his mechanical principles. Two bodies, "M" and "m" distant by "r", attract each other according to the following formula: $f = g(M.m/r^2)$; g is a factor of proportionality, the gravitational constant

The problem that Einstein had with the above description was of two folds: the existence of a force of attraction acting at distance, and instantaneously. It was incompatible with the principle of Special Relativity which advocated that the upper speed limit was the speed of light. A signal having a causal effect (attraction) should not travel faster than light. Instantaneity was excluded.

It took Einstein some six to seven years to solve the problem. In 1916 he finally came out with the theory of *general relativity*. According to this theory, there is no travelling signal between two attracting masses. Gravity is not a force but a distortion, a *curvature* of space-time. This space curvature is created by the presence of matter. In developing the subject, Einstein came out with a mathematical description, the so-called *gravitational field equations*. These are among the most complicated of mathematical formulae, and will not be discussed here. We will, however, review the principle of general relativity and the way Einstein reached his conclusions.

First, he proposes a thought experiment of being inside an elevator in free fall. The speed of descent increases constantly as it is submitted to Earth gravity. You and all objects inside the cabin will fall at the same rate than the lift itself, and will have, therefore, no proper movement in relation to the lift walls. You and the objects inside will float within the cabin in a non-gravity condition throughout the duration of its fall. *The accelerated downfall motion of the lift has in fact cancelled out the earth's gravity* inside the cabin.

Suppose, now, that you take the same cabin away from the earth, away from the sun and all other attracting bodies, far away in the middle of an empty space (such a place may only be found in a thought experiment). In that situation the cabin will be in a state of total rest, with no force acting upon it. You will notice exactly the same conditions inside the cabin: a total absence of gravity. Now, if a force from outside is applied to the cabin, pulling it in one direction that we will call *"up"*, you and all objects inside will be pushed *"down"* to the opposite side of the cabin, *the floor*. A force of gravity is created by this accelerated upward motion; exactly as if you were on the surface of an attracting body like Earth.

From the two experiences above we can then conclude that *accelerated motion mimics completely gravitation*.

Einstein promoted this phenomenon to a major principle in physics: *the principle of equivalence;* equivalence between gravitation and acceleration. This forms the principal basis of the theory of General Relativity.

As we have seen earlier, the central principle of inertia indicates that a *uniform motion in a straight line* is totally equivalent to a *rest situation*. We could now add to this: that an *accelerated motion* is totally equivalent to *gravitation*.

Let us now have a look at an accelerated motion. In physic, an accelerated motion is a motion which *increases its speed* or *changes its direction* continuously. An object in motion, when in proximity and in relation to an attracting body like the earth, will either fall straight to its surface, *increasing* its speed in falling; or turn around it in an orbit, *changing* its direction constantly. Both are accelerated motions. Nowadays there are plenty of examples of the second variety orbiting the earth in the form of satellites or space stations. Men inside it live in a non-gravity condition and float within their capsule just like inside the falling lift. Physically they are unaware of their capsule's movement. For them, the capsule advances with constant speed straight ahead in what appears to be a straight line (no need to maneuver to stay in orbit). But after some hour or two—the duration of a complete revolution around the earth—the satellite surprisingly reaches the same position that it occupied before. What happened? The capsule was advancing in a straight line, but the earth's gravitation has curved the space around it.

The theory advocates that matter bends space.

How does it do it? This is much more complicated, and is not the object of this book—but I will, however, give a very brief outline.

The overall form of any object is defined by its internal structure and the tension within it. In stretching a piece of elastic we increase its internal *tension* and change its form. As we let it go, it will assume its initial size and form, correcting its *deformity*. This is the principle of elasticity. Deformity is the result of internal tension: **D = *f*T** ; with " *f* " being a constant of proportionality, variable with each object. Some structures, like a piece of elastic, have a big *f* value, and are easily deformed—little tension would cause a substantial deformity, while other structures with a smaller *f*, such as wood, iron, or stone, etc... with a smaller *f* are less elastic, more rigid. The space-time continuum is no exception, but it is extremely rigid. Its *f* is extremely small (2.10^{-48}). It would need an enormous tension to induce a small deformity.

Tension is the measure of the quantity of *matter* within the object. The more matter, the higher the tension. To the real conventional *matter*, Einstein, in his gravitational field formula, added two other components: one, the convertible *matter* from energy, using his formula of $E=MC^2$; and two, the *momentum* produced by relativistic movement of weightless particles. The source of *tension* is, therefore, now not only the amount of classic matter, but the totality of *mass-energy*, with those components mentioned above. This forms the right side of the formula (T). On the other hand, for the left side of the formula (D), *deformity* was not easily measurable. It took Einstein several years to express *deformity* mathematically. The challenged demanded the use a very complex mathematical formula, having at least a dozen components. This is why Einstein's gravitational field equations (**D** = *f***T**) are so notoriously complex.

With the above summary description of the theory of General Relativity, we can now draw the fundamental conception:

Mass-energy is the source of *Tension*; and *Tension* causes *Deformity* of *space-time*.

Applied to our earth-satellite system, we should now be able to understand the circular movement of the capsule as it follows the geodesic line of a curved (deformed) space.

With general application of the above concept one can make the following statements:

- The earth and its environmental space make a unit and a continuum entity. The limit of the earth is not its surface, but envelops the space around it: the space that has been shaped by its presence. This is true for all other masses in the universe. The space is fashioned by the presence of matter. There is no absolute space in the absence of matter.

- The earth and all other celestial bodies, due to their internal tension, deform their surrounding space, creating a *curvature* of space-time. Their gravitational field of influence decreases, however, with distance. In a case of several nearby bodies, such as earth, sun, and moon, the effect overlaps and intermixes in the form of a great and complex net.

- The more massive or dense bodies impose greater deformity. A dense neutron star bends the space around it much more than an ordinary, less dense star.

- The universe itself is a finite spherical concept whose curvature is defined by its internal tension: its critical density. Its radius is infinite and therefore its surface flat.

Quantum theory

Isaac Newton thought that light was made of small grains that we now call particles. But many experiments later demonstrated the *interference phenomenon*, and it was concluded that light behaves exactly like a wave. The problem with the wave structure is the fact that a hot body emanating energy in all wavelengths appears to have unlimited energy, as we theoretically can have an infinite range of wavelength, and energy could be dispersed with all those wavelengths.

To solve the problem Max Planck proposed that light, in its energy-carrying capacity, has to act like particle that carries specific amounts of energy, a *quantum*. Each quantum carries a specific amount of energy, linked to its frequency. Quantum physics was born from this founding argument. For a long time the duality nature of light (continuous, and thus a wave, or discontinuous, and thus a particle) was the major subject of scientific debate. At first, a tremendous number of articles and papers were advanced on each side, attempting to prove or disprove the *wave* nature or the *particle* nature of light. Later, when through incontestable experiments it was proved that light could behave in both manners, scientists exerted an enormous effort to unite the two positions and develop a unique explanation. At present the *wave-particle duality* nature of light is well accepted in quantum mechanics.

In the first half of the twentieth century the foundation of quantum mechanics was established, and later enormously developed through the work of eminent scientists such as Max Planck, Albert Einstein, Niels Bohr, Louis de Broglie, Max Born, Paul Dirac, Werner Heisenberg, Wolfgang Pauli, Erwin Schrodinger, and others.

Quantum mechanics is a set of scientific principles governing the *energy* and *matter* of the very small. It examines the rules governing matter and space at the atomic and subatomic scale. Planck's hypothesis states that energy (e) is emitted and absorbed by *quantum*, which is a discrete and a finite "energy element". There can be no fraction of it. A quantum is a ball of energy who's intensity is proportional to its frequency, (v). This is expressed as "$e = h.v$ where h is a factor of proportionality," Planck's constant. Einstein himself postulated that light is composed of individual *quanta*, later called *photons*.

Planck's principle was later applied to all other subatomic elements. The structure of the atom and its electrons' distribution underwent enormous changes both in the perception of structure and in the understanding of behavior. In classic physics the atom was described as being composed of a nucleus around which electrons revolved in a set of well-defined orbital levels. In quantum mechanics, the position of an electron around the nucleus of an atom is not defined at any given time. Its presence at any given position around the nucleus is rather a matter of *probability* at any moment.

We talk about a *quantum field* of electrons. This is true, in fact for all other subatomic elements in nature. This is expressed by the *Heisenberg principle of uncertainty*. This says that the more we know about where a particle is, the less we can know about how fast it is moving. In other words, the more we know about the speed of the elementary small particle, the less we can know about its position. The principle also says that there are many pairs of elementary particles for which we cannot know all about any single one, no matter how hard we try. The more we learn about one of such a pair, the less we can know about the other. In the subatomic world the very act of measuring disrupts the factual situation, so that we could never know both the actual position and the speed of a particle at a given time. If we measure its speed, its position will be among one of a *probability* distribution; and if we locate it precisely, its speed will be within the *probability* distribution; but will not be an exact quantity. This incapacity of exact measuring is not due to the lack of sophistication in our measuring instruments, but is a very way of nature in the elementary world.

No matter how hard we try, classic *determinism* is lost in the context of the subatomic world. Generally, quantum physics does not assign definite values to observables. Instead, it makes predictions using probability distributions; that is, the odds of obtaining possible outcomes from measuring an observable. In the everyday world, it is natural and intuitive to think of everything (every observable) as being really *now* and *there*. Everything appears to have a definite position, a definite momentum, a definite energy, and a definite time of occurrence. However, quantum mechanics does not pinpoint the exact values of a particle for its position and momentum (since they are conjugate pairs) or its energy and time (since they, too, are conjugate pairs). It only provides a range of probabilities of where that particle might be; given its momentum.

The strange probabilistic nature of the quantum realm is not confined to the so-called natural subatomic domain. The annihilation of particle-antiparticles, revealed dramatically by the particle accelerators in the emergence of new and exotic particles, is an excellent example of the fact that the strange implications of quantum theory are more than just a temporary limitation of the mathematical scheme. The "strangeness" is an actual view of a reality that does not conform to everyday, common-sense logic. Virtual particles with negative energy, just like antiparticles with opposite charge, actually can exist, at least for a tiny fraction of a second, as if they were perfect "mirror images" of their normal partners. In particular situations, like the inside a black hole, they can even become *real* particles.

Einstein, being himself one of the founders of the quantum mechanics, is well known for rejecting some of the claims of quantum mechanics. While clearly contributing to the field, he did not accept the more philosophical consequences and interpretations of quantum mechanics, such as the lack of deterministic causality and the assertion that a single subatomic particle can occupy numerous areas of space at one time. This rejection is the source of his famous quote, "God does not play dice with the universe." He produced a number of objections to the theory, which he considered incomplete, and debated with Niels Bohr for a very long time. He was ultimately proved to be wrong.

The theory of general relativity and the quantum theory do not directly contradict each other, at least with regard to their primary claims. They both have definite postulates that are indisputably supported by rigorous and repeated empirical evidence. Yet they are resistant to being incorporated within one cohesive model. Attempts by many distinguished physicists, including Stephen Hawking, to unify quantum mechanics and general relativity into a *theory of everything* have not been successful so far.

Nevertheless, in the light of further developments in more recent years, it has become clear that quantum mechanics is an accurate and valid theory. Application of the principle of quantum mechanics now constitutes the foundation of recent development in computer design and in several other fields demanding micro-techniques.

The duality between the Quantum theory and General Relativity is not solved today, but the *Copenhagen interpretation*, developed largely by the Danish theoretical physicist Niels Bohr and those following, is the interpretation of quantum mechanics most widely accepted amongst physicists these days. According to this interpretation, the probabilistic nature of quantum mechanics predictions cannot be explained in terms of some other deterministic theory, and does not simply reflect our limited knowledge. Quantum mechanics provides probabilistic results because the physical universe is itself probabilistic rather than deterministic.

Chapter II

The Origin of the World

Where does it all come from?

Can something be generated from nothing?

These two fundamental questions have been debated since the earliest history of human reflection, and still remain unanswered.

We will not discuss here the theological, the spiritual, or the philosophical aspects of the questions. This is not because they are less valid, but because they do not reflect the field of our vision in this work, which has a scientific orientation. The area of *belief* is deep and individual, while the area of *knowledge* is common, debatable, and subject to change.

The current answer to the first question, developed in the light of accumulated knowledge in the course of the last century, is obviously "the Big Bang". Nowadays, almost all scientists accept that space, energy, and time appeared all together some 13.7 billion years ago, in a primary event called the BB. This view is not the thesis or the opinion of individual scientists, but has gradually imposed itself on the extended scientific community as a result of accumulating evidence obtained in particular from cosmological observations.

In this chapter we will discuss whether the BB can really be considered as the "origin" of everything, or it is, rather, a step in the evolution of the universe—a step useful to explain the observables.

But before going any further, let us clarify some usual and insistent questions about the BB.

"*What was **before** it?*" is often the first and the inevitable question. The answer is ***not*** "nothing". The question is obviously void of sense, as there was no *before*. Time itself was also initiated with the BB. There is no locale that is north of the North Pole!

"***Why** the BB?*" has no answer either, as it presumes to ask why *something*, instead of *nothing*? Why *being* at all? This fundamental and rather philosophical question introduces guesses, convictions, and beliefs. It remains outside the scope of science, which tries to explain the *how* of things rather than their *why*.

Is the BB really the origin of everything?

In the course of the last century, the universe itself has gradually lost its aura of mystery and has entered the realm of science as almost a physical object liable to observation and explication, and subject to the rules and regulations of physical laws. More than everything else, it is Einstein's theory of Relativity that has contributed to this end. This theory has been the nucleus of formulation for different models of the universe for the last sixty to seventy years. Further development, reinforced specially with Hubble's discoveries, indicated that the universe was not static, but in continuous expansion—a notion that rapidly gained full acceptance.

From there on, it was simple to look backward and run the movie of existence in reverse to reach a state where all existing universe was concentrated in a single point. Applying the same rate of expansion to the past -from the beginning to now, that point is calculated to be at some 13.7 billion years ago. BB was hence created in the scientific language, and has settled in for good. BB is the inevitable, the logical conclusion of the theory of gravitation. It is enforced by the cosmological observations of the last seventy years. BB imposes itself, despite the fact that science is not yet able to describe it fully. Accepting the BB means accepting that the totality of the universe's energy was once concentrated in this dimensionless point which we call a *singularity*.

What is a singularity? And how can we study that state of affaire?

A dimensionless point is already unphysical by itself. When it supposes to contain the totality of the world's energy, it becomes unrealistic. Nevertheless this seems to be really the case. As with anything in physics; there are three ways to study this unique phenomenon. By direct observation, which is not possible; BB is not there for us to see. Simulation in advanced nuclear research centers such as CERN, although promising, falls very much short of the necessary energy intensity for the moment. Only the application of our known scientific laws and theories may bring some hope to shed light on the situation to some degree. Let us try it.

The Planck Era, a lawless land

In Quantum physics, the world of infinitely small is governed by a bunch of laws and regulations called "the standard model of physical particles." which has proved its validity many times over. It concerns the fundamental forces of nature in the realm of subatomic particles: the electromagnetic, the weak, and the strong nuclear binding forces. These are the forces that cause the cohesiveness of the atomic nucleus, and hence determine the very structure of matter in the world. The world of the infinitely small is also determined by Planck sub-atomic values. In the quantum world, the smallest possible unit of energy is Planck's constant (6.622×10^{-34} joule/sec). This corresponds with other inferior limits such as that of length (10^{-35} m), and that of time (10^{-43}sec). There is no going beyond those limits. Beyond these values, it seems that the very nature of space-time changes drastically; time loses its attribute of duration and space its attribute of length (their continuity).

In modern cosmology and BB description, *Planck Era* is precisely applied to a period between instance zero and 10^{-43}s.

As we have seen before, Quantum physics rules the world of infinitely small: small distances, small time scales, and small energy. The forces involved are the three fundamental forces of nature. They define the structure of matter.

On the opposite side of the size spectrum, the force of Gravity, defined by the theory of General Relativity, deals with the world of infinitely great: great distances and great energy. Using the mass-energy concentration, it determines the space-time configuration; and is at the basis of the current conception of the universe.

Those are the two ends of the scale, the two extremes; and the two theories to explain each the infinitely great and the infinitely small. In the actual word, there is no conflict between these two theories, as their fields of action are completely separated.

Or, BB is the unique and exceptional situation that combines the two; an enormous energy concentration in a very small dimension.

Before the conception of the BB, no physical system was known that required the application of the two theories for its explanation and comprehension. The BB is the only situation where there is a need to overlap the two theories. In a singularity, matter and space-time lose their very nature and are intermixed together. In the description of the BB, in order to enter the Planck Era and reach "instance zero," we need to combine the two theories into a *theory of everything*. None of the two theories alone is capable of describing that situation.

In the last three or four decades major efforts have been made to unite the two theories (quantification of the theory of Gravitation) or to develop new ones that would integrate the two. String theory and the theory of Quantum vacuum are among these approaches. Despite promising prospects, none of them seems to be fully successful. These theories often add extra-dimensions to space-time and also introduce many infinite values into the system. The super-complicated mathematical formula, issued from these theories often lead to eliminate simply and squarely "instance zero"; no singularity, which they are supposed to describe.

Out of nothing

Let us now consider the second question.

Can "something" be generated from "nothing"?

The classic scientific answer is obviously negative. But in the world of quantum physics the answer may be different. What is "nothing"? A total and complete vacuum?

Let us imagine that we can completely empty a small portion of space of its contents—every molecule, atom, subatomic element, down to its last photon is aspirated out. Would that be equal to a portion of "pure space"? The answer is negative in both Relativity and Quantum theory. The first would say that such a space does not exist, and the second that it still contains something: energy. According to Quantum field theory, electrons and all other sub-atomic particles do not have a definite and observable position at all times; but their existence in a particular position is a matter of probability. Remember, the modern structure of atom with rather a *cloud* of electrons around it nucleus, with no distinct and observable individual positions. In the ghostly sub-atomic world, *virtual particles* can appear and disappear at all time in what is known as a "quantum field". They do not have enough energy to have a proper physical existence, but may materialize out of a void, borrowing energy for a small fraction of a second before coupling with their anti-particle and being annihilated. The energy comes from the void, the *vacuum energy*, negative and antigravity, but still energy. The energy of a quantum field is never equal to zero everywhere. The "void" is therefore not equal to "nothing". One can theoretically eliminate all physical *matter* from a space, but can never extract its *quantum field*. Space and matter (convertible energy) are therefore inseparable. Absolute space does not exist.

If the void contains the very potential for existence, could the universe have come to a state of being in a process of self-generation?

Can the vacuum energy be considered as the origin of the creation of matter? And in that case, was the BB being nothing but a step in the evolution of the universe after all?

Is a *singularity* the limit between something and nothing? Singularities are also supposed to form the centers of black-holes. Anything that approaches these centers is aspirated in and reaches the end of its physical existence. Light itself is also trapped; these regions are therefore invisible from our space-time. What would an alien see, sitting on the other side of a black hole? Materialization of things from nothing? An act of Creation? A mini BB? Shall we therefore accept not a single universe, but a diversity of many parallel universes—a *multiverse* with a lot of *probability*? Is our universe nothing but one of those probabilities?

The last word is not yet said on the subject of the origin of the world, as the unreachable "instance zero", the Plank Era, remains frustratingly elusive.

Having summarized the fundamental theories involved in the explanation of the BB, and with the above major reservations regarding the origin of the world in mind, let us now attack the very substance of this work.

Chapter III

Big Bang: The Classic Model

The classic hot BB model is based mainly upon the theory of General Relativity, and cosmological observations.

The main assumption made in the model is what is known as the *cosmological principle*. This says that the universe looks approximately the same from any vantage point. More specifically the assumption is, first, that the universe is *homogeneous*—essentially the same everywhere. And second, the universe is *isotropic*—it looks essentially the same in any direction. We hedge a little when we use words like "approximately" and "essentially," because on a small scale the assumptions are not true. Our part of the universe, near a medium-sized star in a rather average spiral galaxy, is certainly not like the interior of a black hole or the empty space between galaxies. However, the assumption is that on very large scales the universe has no preferred location and no preferred direction. .

One consequence of the homogeneity condition is that the big bang event (if such existed, which isn't in fact absolutely required in all versions of the model) didn't happen at one particular point in space. It must have happened "everywhere." And although the universe certainly appears to be expanding away from us in all directions, the same must be true from all other vantage points as well. This is the familiar analogy to spots on an inflating balloon.

This brings us to another assumption: that the universe is not spatially static, but is in fact expanding, This assumption is founded on the observations of galactic redshifts first made by Edwin Hubble, which since his time have been confirmed in many ways. Yet, in some sense, the assumption of an expanding universe remains "just" an assumption. There are still astronomers and cosmologists today who seek alternative explanations for the observed redshifts. If one of these alternatives were correct it might not be necessary to assume the universe is expanding, as it appears to be. However, if the universe is not in fact expanding, then a variety of other things we can observe, such as the cosmic microwave background radiations (CMB), have no obvious explanation. We make the assumption of expansion not only because that's the simplest explanation of Hubble's observations, but also because it yields a consistent model that explains many other things too—such as the apparent age of the oldest stars, the chemical composition of the universe, and the CMB.

Modern cosmology is essentially based upon the theory of general relativity. *Mass-energy* and *Space-time* are unified in one entity. *Gravity*, created by mass-energy, defines the space-time deformity. The universe and its ultimate fate depends upon the mass-energy that it contains; the so-called *critical density*. Higher density will end up with a *closed* universe; and lower density with an *open* one. The greater the tension, the more important is the space-time curvature. At the Big Bang, the deformity was at its utmost value, and exceeded the *elasticity* threshold: a *singularity*. A rupture was produced, in analogy to the rupture of an elastic tape stretched beyond its maximum limit.

It's worth noting that Einstein himself, shortly after developing the general theory of relativity, continued to assume that the universe was not expanding. Before Hubble almost everyone assumed the universe was static. Einstein was, in fact, much annoyed because his theory strongly implied that the universe *shouldn't* be static, and ought to be either expanding or contracting. After all, an object thrown into the air from the surface of the

Earth rises at first, and then it falls back under the force of gravity. It would be very surprising if the universe, under the force of gravity, didn't behave similarly. Einstein even went so far as to modify his theory to allow for the universe to be static by adding the **cosmological constant** (an antigravity force) to it. As it turned out, after Hubble proved the universe wasn't static it seemed that adding the cosmological constant was a mistake.

In this chapter we will detail the two principal arguments in favor of the currently accepted Big Bang model and also the problems presented that argue against it.

Evidence for Big Bang

1- The Redshift

Redshift is defined as an increase in the wavelength of electromagnetic radiation received by a detector, compared with the wavelength emitted by the source.

The Doppler redshift results from the relative motion of the light-emitting object and the observer. If the source of light is moving away from you then the wavelength of the light is stretched out, i.e., the light is shifted toward the red. If the source is moving toward you, the light is blueshifted. These effects, individually called the blueshift and the redshift, are together known as Doppler shifts. The shift in the wavelength is given by a simple formula:

$$Z = (\text{Observed wavelength} - \text{Rest wavelength}) / (\text{Rest wavelength}) = (v/c)$$

This is true as long as the velocity (v) is much less than the speed of light. When velocity is comparable to the speed of light another formula, a relativistic Doppler formula, is required.

Hubble's observations

During the 1920s and '30s, using the above phenomenon, Edwin Hubble discovered that the universe is expanding, with galaxies moving away from each other at a velocity given by an expression known as *Hubble's Law:* $v = H.r$ Here v represents the galaxy's recessional velocity, r is its distance away from us, and H is a constant of proportionality called *Hubble's constant.*

The exact value of the Hubble constant is somewhat uncertain, but is generally believed to be around 70 km per second for every mega parsec in distance (km/sec/Mpc). A megaparsec is the distance covered by light in 3 million years. This means that a galaxy 1 mega parsec away will be moving away from us at a speed of about 70 km/sec; while another galaxy 100 megaparsecs away will be receding at 7000 Km/sec. So essentially, the Hubble constant sets the rate at which the Universe is expanding.

Additionally, the present age of the Universe can be assessed vis-à-vis the Hubble constant: the inverse of the Hubble constant has units of time. By substituting in kilometers for Mpc in the Hubble constant, we find that upon inverting H we get a quantity with units of seconds (kilometers canceling out in the denominator and numerator). For a Hubble constant of 100 km per second per Mpc, we get an age of some 3×10^{17} seconds, or about 10 billion years.

The standard picture of cosmology, based on Einstein's general theory of relativity, explains how to picture this expanding universe. As an example, consider a loaf of bread, with raisins sprinkled evenly throughout it. As the bread expands during cooking all the raisins are moved further and further apart from each other. Seen from the viewpoint of any one raisin, all the other raisins in the bread appear to be receding with same velocity, although each raisin has no proper displacement speed.

This model also explains the linearity of *Hubble's law*— the fact that the recession velocity is proportional to distance. If all the lengths in the universe double in 10 million years then something that was initially 1 megaparsec away from us will end up a further megaparsec away. Something that was 2 megaparsecs away from us will end up a further 2 megaparsecs away. In terms of the speed at which the objects appear to be receding from us, the second object has receded twice as fast!

In physical cosmology Hubble's law is the statement that the redshift in light coming from distant galaxies is proportional to their distance, and denotes their speed of recession. The law was first formulated by Edwin Hubble and Milton Humason in 1929, after nearly a decade of observations. It is considered the first observational basis for the expanding space paradigm and today serves as one of the pieces of evidence most often cited in support of the Big Bang. The most recent calculation of the proportionality constant using 2003 data from the satellite WMAP combined with other astronomical data, yielded a value of 70.1± 1.3 km/s/Mpc. This value agrees well with that of 72 ± 8 km/s/Mpc obtained in 2001 by using data from NASA's Hubble Space Telescope. In August 2006, a less precise figure was obtained independently using data from NASA's orbital equipment.

From this, we can draw a very interesting and rather disquieting conclusion, which leads directly to the "Big Bang" idea. The conclusion is that if we run the evolution of the universe backward there should be a time in the distant past when everything we can now see was a heck of a lot closer to everything else. It would have been *very* crowded at some point in time. While this does not *prove* that at some time long ago everything was bunched together in a state of very high density, it certainly suggests the possibility. Physicists like George Gamow made this and other inferences that eventually led to the theory of the big bang—the notion that at some point in time about 14 billion years ago (by current estimation), all matter in the observable universe was in a state of extremely high density and temperature, and subsequently "exploded," resulting in the expansion we still see today. ("Explosion" is used metaphorically. Exactly what did happen is still a major open question.)

2- The CMBR

Existence of Cosmic Microwave Background Radiation is perhaps the most conclusive piece of evidence for the Big Bang. Discovered almost accidently by A. Penzias and R. Wilson in 1964, it soon became evident that it emanates uniformly from all directions in the sky, and has an almost uniform temperature of 2.70 K. In fact George Gamow and others already had predicted its existence in 1948, while explaining their theory of the evolution of early universe. At first the universe was very hot and dense; radiation and matter were intermixed in the form of plasma. As the universe expanded and cooled, there came a point when radiation decoupled from the matter and cooled gradually to its actual 2.70k. The microwave background radiation and its almost isotropic nature, is now universally accepted as the replica of that decoupling. In fact, the only satisfactory explanation for the existence of CMBR lies in the physics of the early universe. Its incontestable existence has no other current explanation.

Unsolved problems with standard classic model of the Big Bang

The following are some of the problems and questions that the classic BB theory does not explain fully.

1- The age problem, discrepancy

There are two ways to estimate the age of the universe: by studying the life cycle of the oldest stars; and by measuring the rate of expansion of the universe.

The life cycle of a star depends upon its initial mass. The bigger the mass, the brighter is the star and the shorter its life, as it burns its fuel faster. Our sun's estimated life cycle is around 10 billion years; half of which is yet to come. Globular clusters, discovered recently, are dense collections of stars found in far distances from our solar system. Their centers have millions of stars with different size and brightness. The oldest globular clusters contain stars dimmer than our sun. This means that there exist stars out there that have been around at least for some 11 to 18 billion years.

This is inconsistent with the range of age obtained by the second method.

The second method estimates the actual expansion rate of the universe, and extrapolates it into the past until the BB, on the assumption that the rate has always been the same. This is the area Hubble pioneered in the 1920s.

The validity of Hubble's law hinges on the existence of a good and reliable method for determining the distances involved. Measure of cosmological distances is not a simple task. If you know how bright a particular star is intrinsically and you know its apparent brightness as seen through a telescope, then the distance can be computed by the ratio of intrinsic to apparent brightness. This may work for the nearby stars, but for the more distant and faint galaxies this yardstick may not be accurate. The problem lies in being able to determine the intrinsic brightness of a given star. As stars and galaxies come in a wide range of brightness, astronomers must rely on indirect means to estimate the actual luminosity of a particular object. The method is far from accurate.

The first estimations of the age of the universe calculated using Hubble's constant were way off, giving an age range of 3–5 billion years: A universe much younger than some of its constituent parts! For many years there was a serious problem with that interpretation. It turns out that, because of the great difficulty of measuring cosmic distances, Hubble initially underestimated distances. Since there was little uncertainty about the redshift, and hence the recession velocity, this made the "age" of the universe a mere 2 to 3 billion years. Other means of estimation of the age of the solar system and the Earth pointed to an age of 4 to 5 billion years, let alone the ages of older clusters, estimated at 11 to 18 billion years as we have seen above.

Recently with improved measurement of distances, the discrepancy is no longer so flagrant, but the crisis persists.

2-The flatness problem

The shape of the universe is determined by the tussle between the force of gravity and the momentum of expansion. The strength of gravity depends on the density and internal pressure created by the matter content of the universe, denoted by Ω (omega). The curvature is positive, zero, or negative according as omega is greater than 1, exactly 1, or less than 1. Measurement of the density in the last twenty years indicates an Ω range of 1($\pm$01). However, this is a serious problem, because the Friedmann equation predicted that because the universe had expanded so much since the big bang that Ω would need to have been extremely close to 1 during the earliest moments. Such a cosmological fine-tuning seems to be an extraordinary coincidence. This is called the **flatness problem**; since $\Omega \approx 1$ is equivalent to having zero curvature—indicating a completely flat universe, in other words. Although the estimated value of Ω hovers around 1 these days, the range implies a lot of uncertainty. Part of the problem lies in the fact that we can only see a small part of the matter (10-15%). The rest is inferred from the gravitational motions of galaxies and other observations. The existence of a form of matter with strong negative pressure, "dark energy", is now accepted. It is called the cosmological constant: a curious, antigravitational force that is supposed to increase its grip as the universe expands.

3- The horizon problem; lack of causal connection

The extraordinary homogeneity of the universe forms a major problem in the standard BB model. The universe is, in fact, extremely homogeneous, confirmed by the spectrum of the microwave background radiations everywhere in the universe. The problem is the almost unique temperature, of the universe, 2.7K, (the Bolzmann constant), being the same anywhere; a problem of connection. How has this homogeneity come about?

The reason is that parts of the universe that we can now observe on opposite sides of the sky were, during the first instants after the big bang and ever since, too far apart from each other for light or for any other kind of information to travel between them. Therefore, it is very hard to understand how different parts of the universe that were not "causally connected" at very early times could have turned out to have identical statistical properties, such as the same temperature, for example. This puzzle is known as the *horizon problem*, because at very early times most parts of the universe would have been "beyond the horizon" from each other. The light-travel time between any two points exceeds the age of the universe. Consider this: When we look deep into the sky, distances also correspond to time into the past. If we look at the opposite sides of the universe—say right then left, or up and down—with more and more sophisticated instruments (Hubble's telescope), we look deeper and deeper into the past. The deepest possible vision is near the BB, some 14 billion years ago. Light reaching us from the left has taken 14 billion years; light coming from the right also 14 billion years. There is just not enough time from the BB until now for the light to have gone from left to right! It would need two times 14 billion years. With the actual accepted speed of light, no causal communication between different parts of the universe could have been possible at any time during the evolution of the universe. How can we then explain the incredible homogeneity?

Theory of inflation

In order to overcome the above problems, Alan Guth and others have advanced the theory of inflation, which has gradually won by default. It proposes a period of extremely rapid expansion of the universe in its early times. According to this theory, the dimensions of the early universe increased by a factor of at least 10^{26} in a fraction of a second. While the cause and detailed mechanism responsible for this sudden and extremely short period of expansion is not known, negative-pressure vacuum energy is advocated as the possible explanation. Many questions remain however unanswered. Why would there be that huge negative pressure energy (cosmological constant)? And why would it act only for such a brief period? Where did it come from? How does it decay afterward? What does it mean to postulate a space expansion with nothing in it? Has absolute space any sense? And if not, what was distributed in it?

A new concept of the universe

Everything is relative, even the speed of light.

In our conceptual vision of the BB, as we shall see, none of the above problems are encountered, as the constancy and uniqueness of the speed of light is not assumed in cosmological terms (see the section on redshift). We advocate that a Primary Message was delivered at BB, which defined the fundamental features of the universe for good. A message disseminated at the *then current* speed of light—a speed quite different from that we recognize these days. If the fact of the homogeneity of the universe is correct, as it seems now to be definitely the case, the assumption of such a super-fast primary message becomes a logical conclusion. In the very early moments of the universe it is the beam of light that has delivered the causal message, homogenizing the universe. But for that, light must have had a much higher speed to be able to deliver the message to everywhere; the actual speed of light would not have been sufficient (*horizon problem*). Even if a period of fast universal expansion is considered at the outset, pure space expansion is inconceivable. It is our opinion that something moving at a much higher speed than that of the actual light speed (may be photons or virtual photons) was distributed during that expansion; a speed that decreased with time and distance according to a particular and precise gradient. We advocate that this was the message that defined the fundamental features of space-time in the universe for good (see thesis postulate).

Chapter IV

Big Bang: A new Model
Our Conception

Here also, the model is based upon the theory of relativity, of gravitation, the Hubble observations, and quantum theory.

As the principle of *escape velocity* serves as the basis of our proposed new model, we briefly describe escape velocity in classic physics.

Escape velocity

In a given gravitational field, escape velocity is the speed required for an object to break free from the attracting source and never return to it. The weight of the object or the direction of its movement is irrelevant, only the speed counts.

Figure 1: Newton's canon

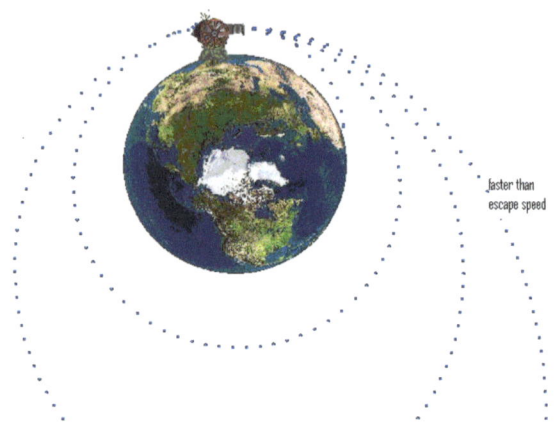

In the picture above, Isaac Newton's imaginary gun is shown shooting a projectile horizontally from a high mountain. As the speed of shooting increases the projectile lands further and further from its source. At a certain speed the projectile will not fall to the ground, but will assume an orbital movement around the earth. The orbiting satellite is a common notion now.

The height of the orbit and its form depend on the initial speed. In its ascending course, the speed of the projectile decreases as it gets farther from the source and this continues until it reaches its maximum height. In the return course, the speed increases accordingly. This is due to the fact that the *kinetic energy* must be equal in magnitude to the *gravitational potential energy* at all times.

As the launch speed increases, there will be a moment when the *escape speed* will be crossed and the projectile will leave the earth attraction for good and move away forever, until attracted by other celestial body.

Somewhere in between, at the precise critical speed—that of escape speed—*theoretically* the object will not really leave the earth attraction. But while reducing constantly its speed, it will continue its course to infinity where it will "hang" with zero speed, forever.

It will not return and it will not move away.

In that case, the speed assumed by the object at any given position along its trajectory is the precise escape speed at that point.

Defined a little more formally, in a gravitational field escape velocity is the initial speed required for an object in any point in the field to reach infinity with a residual velocity of zero, relative to the field. Conversely, an object starting at rest and at infinity, dropping toward the attracting mass, would, throughout his trajectory, move at a speed equal to the escape velocity corresponding to its position.

Thus any gravitational field has its escape velocity.

Some Examples

Earth

On the surface of the Earth, the escape velocity is about 11.2 kilometers per second

However, at 9,000 km altitude in space, it is slightly less than 7.1 km/s.

Sun

The sun's escape velocity at its surface is estimated at 617.5 km/sec.

On the earth, because of its distance from the sun, the escape speed in respect to the sun gravity is reduced to 42.1 km/sec. To this one has to add Earth's own (11 km/s).

Our galaxy, the Milky Way

The Sun is situated in one of the galaxy's wings far away from its center. In this region of the galaxy (our solar system and its dependents), the escape velocity with respect to our galaxy's gravity is estimated at some 1000 km/sec.

The escape velocity near the center of the Milky Way must, therefore, be much higher.

Multiple sources

The escape velocity from a position in a field with multiple sources is derived from the total potential energy per kg at that position, relative to infinity. The potential energies for all sources can simply be added. For the escape velocity this results in the square root of the sum of the squares of the escape velocities of all sources separately.

Beyond the Milky Way

To leave the Milky Way a celestial body must acquire and exceed the galaxy's escape speed. In that case it will continue its course into the depth of space until attracted by other galaxies or clusters of galaxies.

And the universe:

If we consider the whole of the universe as a unique gravitational field, a system of escape velocity may be attributed to it.

Thesis postulate

The universe as a whole with its gravitational field is taken as a unit. In accordance with classic physics, a system of escape velocity may be attributed to this unit.

We will call it *universal escape speed*; a new notion. This is the ultimate upper speed limit in the whole universe, past, present, and future. Going faster would simply mean "leaving" the universe; an obvious absurdity.

In physics the highest known speed is that of light. Nothing can go faster. We postulate that the "universal escape speed" and the "speed of light" are equivalent, and are the same phenomenon.

Thus the speed of light as we know it (300,000km/s) is not just an inexplicable physical constant; but is *determined* by our time-location in the universe. We postulate that light propagated much faster in the distant past, and that its speed is different in other regions and epochs in the universe. Light's speed remains, however, now as well as in the past, *punctually invariable* and always an upper limit for that environment.

In fact, it is the rate of the flow of time that is different cosmologically (see section on Time); but for easier comprehension we will continue talking about light-speed variation through this book.

The proposed model of the universe

It is based on the principle of Universal Escape Speed.

We will discuss it under the following headings: *the Big Bang, the Redshift,* and *the Phase of Continuous Expansion.* Other chapters (*Soon after the BB* and *All in the Middle*) will follow. The last chapter on the *Nature of Time* will complete our conceptual vision.

The Big Bang

Two assumptions

First, it is essential to notice that the homogeneity of the universe is *not* assumed *with respect to time*, even though space-time is still assumed to be four-dimensional just as described in special relativity. The symmetries pertain strictly to the three spatial dimensions. It is in fact definitely assumed that the universe must have been very different at times sufficiently far back in the past, as well as (probably) far in the future. The so-called "steady state" cosmological model that assumes time symmetry—that the universe has always looked about the same and always will—seems to conflict in a number of ways with what we actually see around us. For instance, there are several compelling reasons to believe that the universe must have been much hotter and denser in the past than it is now.

The Primary Bang

At the very beginning, cosmology and high-energy particle physics overlapped completely. If you recall, one of the basic ideas of the standard model of particle physics is that there are three fundamental forces in nature: the strong nuclear force, the weak atomic force, and the electromagnetic force. The force of gravity, related to matter, is the fourth factor. It affects large objects and has practically no role in the subatomic elements. However, at the very earliest time in the existence of the universe, all those four forces were "unified" and indistinguishable.

At the primary instance, the BB singularity, there is no means of understanding or explaining anything, as there is no known law in physics to explain that particular situation. Gravity is not explicable by the general theory of relativity, and the quantum theory is not applicable, as the energies that are involved are enormous.

Past the *Planck Era* (10^{-43} sec), the contemporary model advocates the phenomenon of *Inflation* in order to explain the incredible homogeneity of the universe (*horizon problem*). Although an adequate theory to explain or describe this process does not exist, the classic model advocates that a very abrupt change in the universe occurred, in which the size of the universe increased dramatically. This is the famous theory of inflation, which is actually the only way in the classic model to explain the causality effect for the universe's homogeneity.

The problem is that it involves space expanding at a tremendous speed, much higher than the actual speed of light. What were the reasons for that sudden change in the rate of expansion?

Even if we accept the theory of inflation, we still face the question of what was distributed in that primary wave of expansion. Space expansion alone suggests *absolute* space; a notion not accepted these days. What then?

Basic to all physical theories are certain fundamental quantities: matter, length, and energy. Matter could not be distributed with that speed. Expansion of space with nothing in it is meaningless; it would be undistinguishable from what comes beyond it; the notion of absolute space with nothing in it is inconceivable. So the elementary answer to the question seems to be that energy distribution accompanied space expansion. Photons or a virtual component of them were distributed in a form of a message at a much higher speed than that of our actual light speed.

We have to consider that, past those very early instants, the universe was in a state of near infinite density and temperature. The almost infinite curvature of the space was created by almost infinite gravitational field. Nothing could escape that tremendous gravitational pull, as the force of gravity had now become operational and distinct from the other three forces. In the language of the General Relativity, we are at the edge of "elasticity" and rupture. Even the weightless photon, probably the only element present at that time, had to acquire an almost infinite, but very specific speed in order to escape: *the universal escape speed*.

It is our concept that this Primordial Message, spread at the universal escape speed, shaped the fundamental structure of space-time continuum for good all along its trajectory. It corresponds to the layout of the laws of gravitation.

We do not suggest here that the expansion rate of the proposed big bang model followed exactly the formulae of escape velocity principle; but the analogy is reasonable. The speed of the primary deployment from an almost infinite density (very small radius, r) and an enormous gravitational constant (G) must have been so fast as to give the look of a "creation". This primary lay-out speed has, however, decreased as time and distances increased from the center. Remember Newton's cannon shooting and its precise critical escape speed. The rate of decreasing speed may not have followed exactly the actual formulae of escape velocity as we know it; but the principle must logically hold.

The model suggests that the Primary Event must have been so finely tuned that its expansion rate corresponds exactly to the *universal es-*

cape velocity all along its deployment trajectory. Gradually decreasing, it reaches zero at infinity, defining the confines of the universe; and—an extremely important fact—resulting in a *flat universe*, the currently accepted version. A lower rate would mean re-collapsing, faster would mean "leaving" the universe.

The Event is obviously unique. It would better be called **Primary Bang (PB)**. It defines the form and fate of the universe at once. Asking why or how that degree of precision would simply mean why the universe and how the BB.

Therefore, as in the classic model, here also the BB (PB) is considered at the origin of the sudden creation of space, time, matter and energy. We advocate that the primary lay-out has left its foot print in the form of an upper speed limit, variable according to its locus, all along its trajectory. It has therefore defined the fundamental feature of space-time relationship anywhere in the universe, far back in the past as well as far in future.

It is clear that during any period in the evolution of the universe, and therefore in any locus in the universe thereafter, this speed remains a non-crossable speed limit—that of *its* light. This speed limit will therefore apply to all subsequent constituents of the universe: stars, galaxies, and their dependents.

The determinant feature is, therefore, the universal escape speed, which is considered here as the speed of light; variable in terms of cosmological time and defined at the outset. For our epoch and our locus in the universe, and with our measuring tools, we have defined it as 300,000 km/sec.

This forms the fundamental basis of our hypothesis. The question of how or why this speed has come around seems irrelevant, as it would involve questioning how or why the BB—a different area of conjecture

The **second** assumption is that after the Primary Event, the universe has not expanded in a constant rate; having expanded at a much faster rate at first, it has decreased its rate of expansion enormously and continuously.

After this primary expansion, and probably under the influence of its impetus, the universe is continuing to expand at a much lower rate. This is the phase of expansion corresponding to the famous "raisin loaf".

Space is expanding continuously with time, giving the appearance of dispersing stars and galaxies; hence, diluting the universe' density. (See chapter on continuous expansion…)

Let us look at the picture:

Figure 2: Different horizons

The concentrated spheres show the universe's evolution since the BB. Each sphere corresponds to a *locus-epoch*; by analogy, to space-time. We will call it a *horizon*. Each *horizon* has its proper speed limit, corresponding to the primary deployment escape speed at that region. Galaxies, stars, planets, and their dependents are born and die constantly in different regions and horizons all the time. All constituents of that region, galaxies, stars, moving objects, and elementary particles, have to obey the speed limit rules of the region, as this corresponds to their light speed.

It is clear that dividing the universe's evolution into separate zones on the picture is to make the point; the reality being a continuum in which density, speed limit, temperature and energy are decreasing gradually.

The Earth, solar system, and the Milky Way are situated somewhere among those horizons in the evolution of the universe.

Let us take out *our horizon* from the assemblage and look at it separately (recognizable by its color).

Figure 3: Our horizon

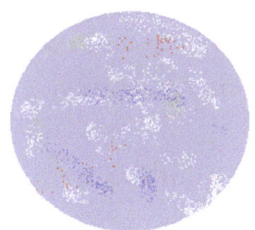

It corresponds to a sphere at the surface of which visible stars and galaxies are disseminated. Light propagates at some fixed 300,000km/s everywhere on this surface, which is the upper speed limit within this *horizon*. In other word if we stick to the unique speed of light (300, 000km/s), the conceptual universe would be the surface of that sphere situated at some 13.7 billion years from the BB, as it is commonly calculated these days.

It is clear that within this concept and the continuous expansion of that sphere we join the classic BB model, where all basic cosmological data holds true. Light coming at a fixed speed from the stars situated on the surface of this sphere show a *redshift* interpreted rightly by Hubble as their recession. The more distant stars have a larger shift and recede, therefore, faster as the sphere blows up. The *Hubble constant* indicates the rate of expansion and is at the basis of our calculation of the age of the universe.

Hence, the accepted classic model confines the universe to this sphere and this single cross-section horizon; a restricted view. It does not attempt to describe those regions of space significantly beneath or beyond it, where space-time features may well be quite different.

The redshift

An alternative explanation

The redshift, defined as an *increase in the wavelength* of electromagnetic radiation, has always been explained in cosmology by the relative motion of the emitting and receiving sources.

In our model the universe is not limited to the 13.7 billion-year old horizon. The model pretends the existence of a continuum, formed at the outset, by a multitude of *horizons*, outer and inner to ours; all inhabited by stars and galaxies. The fundamental feature of the space-time relationship at each horizon is determined by the Primary Event and the principle of escape velocity. This has fixed a specific speed limit for each horizon; that of *its* light, gradually decreasing from the center outward.

It is clear that within this continuum, everywhere in those horizons, stars and galaxies are formed constantly, live their lives, and die all the time. This is probably true everywhere except in the "early days" after the BB and very far near the "border" where special conditions prevail.

Light from other horizons

We have seen that light coming from stars on the surface of *our horizon* reaches us at a fixed 300,000km/s. The redshift observed at the reception undoubtedly denotes the recession of the source compatible with the expansion.

But what about light and information coming from others *horizons*?

Light from stars situated in the outer and inner zones beyond our horizon also reaches us continuously. In fact it is impossible for us to make the difference between the origins of the incoming light arriving from different horizons. These lights may also show a Doppler change, as usual. Their wavelength may be increased or decreased as seen by us at the reception, compared to the emission at their source.

During 1920s and '30s E. Hubble and others observed the redshift from near and distant stars. The redshift, being proportional to the source's recession speed, indicated the expansion of the universe.

Ever since Hubble, and through the last seventy years or so, the scientific community has definitely accepted that light's redshift can only be explained by recession of the emitting source. In fact, no other explanation can be given…so long as we consider the speed of light unique.

Introduced by Einstein and accepted unconditionally ever since, the constancy of the speed of light has never been challenged in the spirit of any scientist; so much so that as an axiomatic dogma in physics, it would never cross anyone's mind to advance any explanation about anything in physics that would cast a doubt on it.

The notion of variable speed of light has been introduced many times in the past without ever gaining any general acceptance in the scientific fields.

What we advocate here is a totally different view on the variability of the speed of light. The concept is based upon the variable rate of the flow of time itself from the BB onward; a rate linked to the general expansion of the universe. The rate has been decreasing ever since. Time itself is, therefore, slowing down, explaining the variability of the speed of light (see chapter on Time). For easier comprehension, we will however continue talking about variable speed of light throughout this book.

In this conceptual vision of a universe with a cosmologically variable speed of light, we advocate that light coming from the *outer* horizons started with a lower speed from its source and has gradually increased its speed to reach us at our 300,000km/s. The light coming from the *inner* horizons has started with a higher speed from its source, and has gradually decreased its speed to reach us, again at our 300,000km/s.

The fundamental point is that in both cases the speed is an *accelerated motion* (positive or negative), and not a linear fixed value. The speed of any electromagnetic wave being the product of its frequency and its wavelength, any acceleration or deceleration in speed will affect the wavelength.

The incoming light reaching us from stars in the outer horizons propagates therefore with a positive *accelerated* motion, and light from the inner horizons with a negative accelerated motion. **We advocate that the observed wavelength changes may also be due, at least in part, to those speed variations.** In fact a decelerating incoming message from a fixed source is equivalent to a uniform-motion message coming from a receding source. In the case where the distance between the emitting and receiving source is fixed, the redshift is proportional to the value of the acceleration. As the acquired acceleration is also proportional to the covered distance, the shift is proportional to the distance between the emitting source and the receiving end. In this view, the bigger the redshift, the *farther away* is the source (the *faster the receding source*, as advocated by Hubble).

In reality and in an expanding universe all constituents participate in the general expansion, moving away each from other. And therefore, according to our concept, the observed wavelength changes are a combination of the two phenomena.

This viewpoint has never been considered in physics because of the unchallenged understanding of the constancy and uniqueness of the speed of light. Any redshift has always been interpreted as coming from a receding source. It could not be otherwise. Light speed was never thought of accelerating or decelerating; and in that case there was no other explanation.

As mentioned above, it is important to note immediately that the stars are not immobile and participate fully in the general expansion of the universe. The observed redshift has, therefore a double explanation: both the receding source and the accelerated light motion.

For example, if the Hydrogen absorption lines measured normally at 4861 and 6563 angstroms are found to be at 4923 and 6647 angstroms in the spectrum of a particular galaxy, the redshift calculated would be : $z = 62/4861$ and $84/6563 = 0.01275$. This, in the Hubble interpretation would mean that that particular galaxy is moving away with a speed of 3826km/s (300, 000 x 0.01275). On the other hand, in a model based on the assumption of a static universe and accelerating light speed, this would mean that the light at the emitting source had a speed of 303,828 km/s (300,000 + 3826), and that the observed redshift was entirely due to the negative acceleration. In fact in our view none of the above interpretation is correct. The emitting light sources are moving away in an expanding universe, but also the incoming lights from other horizons sustain acceleration proportional to the source's distance. Both of these two phenomena contribute to the observed wavelength shift.

Age of the universe

In the classic Hubble interpretation, the totality of the redshift effect is taken as the rate of universe's expansion. *Hubble's Constant* is a common measure used to calculate the age of the universe. As we have seen, this has given rise to interminable debates about the timing of the BB and the age of the universe. A value of 80 to 100 of the Hubble Constant as calculated at first would give an age estimation of some 3–10 billion years to the universe, younger than some of its constituent parts! The more recently refined measure of the Hubble Constant advanced by NASA is 74; giving an age estimate of some 13.4 billion years. The problem, however, is far from being solved.

Two points to consider:

1. As discussed above, the observed redshift in our view is believed to be due to two phenomena; the receding source and the effect of acceleration. This added factor tends to reduce the rate of general expansion, and therefore helps the age problem debate. In order to measure the exact difference, one would have to know the real change in *our speed of light* over a long period of time, say the last few hundred or thousand years. As these durations are trivial in the face of cosmological scale, and as, on the other hand, the measurement of the speed of light with that degree of precision is in the range of measure error, the hope of arriving at a solution seems very remote. Nevertheless it is clear that the option introduces a diluting factor in the observed redshift calculations, helping to resolve the age dilemma.

2. To consider on a more profound level, let us evaluate our system of calculation of the age of the universe. We creatures on the earth, in our insignificant part of a galaxy—one among milliards others—have discovered that light speed is an upper limit of all moving objects. We have also found that our earth is moving around our sun. Its complete revolution, divided by days, hours, and minutes, gives us our *time unit*. We further defined the unit of our length, the *meter*, and we preserve the master meter-measure in a safe place. Using these "on-board tools and data" that serve us in our region, we have very scientifically calculated the speed of light at some 300,000km/s, the speed limit for us. We have then assumed that this speed limit is unique and *permanent* and that it has always been the same from the time of the BB. We have looked at the sky and observed the lights coming from far away stars and discovered in them a *redshift*. In our concept of the fixed speed of light, the redshift could not be due to anything other than the receding of the observed stars. Putting all data points together, we have concluded that, despite many counter-opinions in the beginning, the universe is expanding. We have then run the movie backward, *applying the same expansion rate* to the past, present, and future and we have arrived at a calculation of the age of the universe.

This line of thinking has given rise to many questions and debates over the age discrepancy between the universe and its constituent parts, over information distribution (causal effect) with this rather low light speed at the beginning (horizon problem), over the flatness and dark matter problems, and more. Attempts to overcome these difficulties have either not been generally accepted, have been incomplete, or have won by default (theory of inflation). In any case, in most of them a period of far faster universal expansion, equivalent to a much higher speed of light, has to be assumed.

In our conceptual vision, the universe is not confined to a limit defined by our local speed of light. This is local and today! The universe in our concept is the total existence of space, matter, and energy far back from the Primary Event to well ahead into future. There are epochs in the past where light had much higher speeds than today, and also epochs in the future (our past and future) where the speed is less. Those epochs are also parts of our conceptual universe... Its limit is defined by the gradient of the speed of light, the so-called *universal escape speed*.

With the speed of light cosmologically variable from the BB to now, our conception does away with the notion of time and age altogether for the universe. BB is happening always, even now! Past, present, and future have no meaning as far as the universe is concerned. The *flow of time* imposed upon its constituents—galaxies, stars, planets, and other beings—during their lifespan—denotes the process of aging for *the constituents only*. The universe itself is timeless. This continuous, irreversible flow that we call time is in fact the unperceivable, gradual general expansion of the universe. Its rate has been variable from the BB. Extremely fast at first, it is continuously decreasing. This means a slowing of the flow of time itself (see chapter on Time), which is translated into an apparent gradual decrease in the local speed of light.

Can we feel the gradual change in the speed of light?

At first we might be tempted to think that a world in which the speed of light was slower or higher would be a different world. This would be a mistake. If the cosmological constants: c (light speed), h (Planck constant), and e (energy) were all changed so that the values they have in metric (or any other) units were different when we look them up in our tables of physical constants, but that the value of alpha (α) remained the same, this new world would be *observationally indistinguishable* from our world.

Phase of continuous expansion

After the Primary lay-out and under its impetus the universe continues to expand at a much slower rate. It is generally accepted that in this continuous expansion the constituents of the universe—stars and galaxies—do not have proper movements, but that space itself is increasing, spreading them apart. This is inevitably associated with a decrease in density and temperature.

It is interesting to notice that the universe's expansion became first evident after Einstein introduced his theory of general relativity. When the Russian scientist Friedmann applied Einstein's field equation formulae to the universe as a whole, he obtained results indicating an expanding universe long before Hubble's observations. The finding annoyed Einstein intensely, as he believed, like all of the scientific community of the time, in a static universe. During the early days after the introduction of the theory of general relativity, Einstein had introduced a lambda force, *the cosmological constant*, in order to stabilize the universe, which would otherwise seem certain to collapse due to the gravitational forces. This constant was the energy of empty space, a repulsive force that was supposed to counterbalance the g-forces.

This somehow *static model* of the cosmos, attracted by gravitation force and expanded by the lambda repulsive force, appeared to be very unstable. Lemaitre in Europe and Friedmann in Russia, making full use of Einstein's gravitational field equations, came up with the solution of an expanding universe. In 1926 to 1927, using Hubble's observations, it became increasingly clear that the universe is not static and is expanding. When Einstein finally woke up to these facts, he publicly abandoned his cherished static model in utter disgust as he realized that he could have predicted Hubble's theory of expansion of the universe from the beginning. It was a terrible missed opportunity. With the furious rejection of the static model, out went also the lambda cosmological force all together. True, the *cosmological constant* was no longer required as the universe was considered no longer static, but the fact that the universe is expanding does not necessarily rule out the existence of a lambda force.

Form and fate of the universe

It is interesting to notice that, in almost all of the more recent works, a lambda force has reappeared from total disrepute and is integrated in the explanation of the cosmos's stability, as we shall see.

Lemaitre showed that Einstein's field equations were consistent with a variety of expanding cosmological models starting from a BB. Freidmann had also come to the same conclusion a few years before.

Without going into details, we will describe the current accepted model of Lemaitre-de Sitter.

The fate of the universe in its continuous expansion is determined by the struggle between the forces of gravity, which tries to pull the retreating galaxies back, and the impetus of the BB, helped by the lambda force pushing for expansion. As we know, the force of gravity weakens with distance, and therefore the more the universe expands, the less the breaking force there is. Gravity is created by matter. If there is enough matter (visible and mainly invisible) in the world, the combined gravitational attraction will eventually halt the expansion completely and turn it into a collapse: a big crunch (*Closed model).* On the other hand, if there is less matter, the battle will be won by expansion, and the universe will free itself from all restriction and expand indefinitely in a constantly increasing rate (*Open model*).

At first, it seems that in either case there must be sometime a "moment of truth" at which the outcome will be decided.

Before going further, let us evaluate the opposing forces:

1- Attraction by gravity: This is a direct result of the quantity of matter in the universe. *The visible matter*, galaxies and stars (0.5%), neutrinos (0.3%), heavy elements (0.03%), free hydrogen and helium (4%), form together a minor proportion of the estimated matter in the universe. The bulk of the matter in the universe is in fact constituted by the *dark matter*, with a not well-defined composition. An enormous quantity of matter forming the center of galaxies and maybe clusters of galaxies, in the form of gigantic black holes, together with all other smaller black holes and other dense stars probably forms a part of the dark matter. Matter exists also in the form of energy ($E=MC^2$).

By definition, "Matter" in physics is a collection of primary particles that possesses a rest-mass, such as baryons, leptons, and some neutrinos. In contrast, photons and other particles that move at relativistic speed are called "radiation." They have no rest-mass, but their speed confers to them a *momentum*, creating the all-important pressure (p) within the universe. The totality of the list of elements mentioned above represents, therefore, the combined density of matter and energy and constitutes the force of gravity. The proportion of matter and energy is a subject of debate, and has probably been variable throughout universe's development. But in any case, in a matter-dominated universe or in an energy dominated one, the overall force of gravity is attractive. To fix the idea, the actual density Ω of the universe is defined, with a critical value of some 6.10^{-27} kg/m^3, equivalent of two hydrogen atoms per cubic meter.

2-Repulsion by cosmological constant: This is the famous lambda force (λ), the *dark energy,* which acts as a repulsive force—as an antigravity. This is the force that was first introduced by Einstein to stabilize his universe. It manifests itself across enormous cosmological distances. It is unusual in that it grows *stronger* with distance; in total opposition to all other forces of nature. Lambda is an extremely weak force in ordinary circumstances, so that most of physicists would like to set it squarely at zero. The problem is that in all physical theories that attempt to unify the nuclear and electromagnetic forces of nature, a lambda force keeps cropping up.

Although nuclear and electromagnetic forces have nothing to do with gravitation, a cosmic force mimicking lambda appears as an inescapable by-product. Some recent data coming from measurements of the *cosmic microwave background* radiation *(CMB)* also forces the inclusion of a lambda force. Worse, the magnitude of lambda is unacceptably elevated when cosmologically long distances are considered. In the case of a positive cosmological lambda and in the presence of a constant speed of light, the expansion of the universe would be exponential. It would accelerate the expansion dramatically to blow the universe apart in a few microseconds.

With the above evaluation of forces let us now consider the theoretical outcome.

Almost all the classic cosmological models of the universe now include a lambda factor these days. Being weak at short distances, the lambda force makes virtually no difference in the compressed early stages of the universe. However, as the universe expands, and in so doing decreases its density, the repulsion gains in strength and counteracts the normal action of gravity.

In the Lemaitre-de Sitter model, in the beginning, the gravitational breaking effect dominates and enormously decelerates the primary expansion. But as the universe grows larger, the opposing forces become more evenly matched for a much longer time. A long period of stabilization settles in, during which expansion is slow. This period of a "loitering universe," associated with a lower value of the Hubble Constant (72 instead of 80 and more) is advanced specifically to solve the age inconsistencies. However, this only postpones the issue, but does not change it. Continued expansion will eventually give the repulsion the upper hand, the lambda force remaining constant in the face of diminishing attraction. Slowly but surely the universe will start to accelerate in its expansion, with a lambda force gaining in strength in proportion to attraction. The situation that pertains near the BB is, therefore, reversed. With the attracting force fading away, the universe will blow itself apart (*open model*).

On the other hand, in models in which a lambda force is not included, the fate is even more quickly settled. The attraction of gravity takes the upper hand pretty soon, and a final "crunch" is inevitable (*closed model*). In these models, the age issue of the universe and the lack of time for the formation of stars and galaxies becomes an embarrassing problem with no reasonable answer.

The scientific opinion that finds increasing acceptance these days is neither of the above, but is instead a *flat model* for the universe. A flat universe corresponds to the surface of a sphere whose radius tends toward infinite.

Our view; a flat universe

In our conceptual model presented here, the fundamental consideration is the presence of a *universal escape speed* which has determined timespace features of the universe from the outset. It is considered to be the local light speed. Being highest soon after the BB, it decreases gradually and tends towards zero at infinity, the universe's border. The universe has been laid out according to this speed gradient.

The phase of subsequent and continuous expansion is governed by the action of forces in presence as described above. Galaxies and stars form and die all the time, except possibly in a period soon after the BB and late toward the border of the universe, where special conditions prevail. The general continuous expansion of the universe takes its constituents—stars and galaxies—*across the gradient* of light speed. This fixes their upper speed limit at each point. We advocate that this unperceivable move is translated into the *flow of time* for the constituents of the universe (see chapter on time). It introduces a temporal dimension and a lifespan for stars and galaxies and all other matter and beings. The primordial point is that this temporal dimension does not apply to the universe itself as a whole. The speed of the Primary lay out was (is) that of light speed all along its trajectory. Despite its variability, light speed induces an infinite time warp, stopping its flow. The universe itself is, therefore, timeless.

The other crucial point is that the gradient speed of light associated with the general expansion also influences many other parameters, including the lambda force. In fact the Lambda force being indirectly a derivative of the speed of light, its strength is related to the speed of light. When this speed is taken as a constant in the presence of a decreasing density Ω due to continuous expansion, the lambda force becomes unacceptably high in magnitude. But with a variable speed of light the lambda force follows the proportionally decreasing local light speed.

With this new parameter in mind, let us have a look at our two scenarios of "recollapse" and "free expansion". In the case of a recollapse, the moment of truth would be when the universe reaches its maximum size. In the event of outburst expansion, it would be when the braking effect becomes negligible, and the lambda force gets out of hand.

In either of the two scenarios, the existence of a particular "moment of truth" introduces a notion of time scale into the description of the universe; the time of decision, a time-distance from the BB. Or, in our conceptual vision, as we have just seen it above, the universe itself is timeless, and no such moment could be defined.

In our view, as the gravitational pull declines inexorably with general expansion the lambda is also declining proportionally, as it is tied up to the speed of light, which is now also on the decrease.

As g declines, λ declines also. The tussle is evenly matched. The battle will never end, meaning that the universe will expand forever, but at an ever diminishing rate *(Flat model at infinity)*.

The "moment of truth" is postponed indefinitely in a timeless universe.

There is no time scale or length scale built into our conceptual vision for the universe itself.

See the figure below.

Figure 4: Evolution of the universe

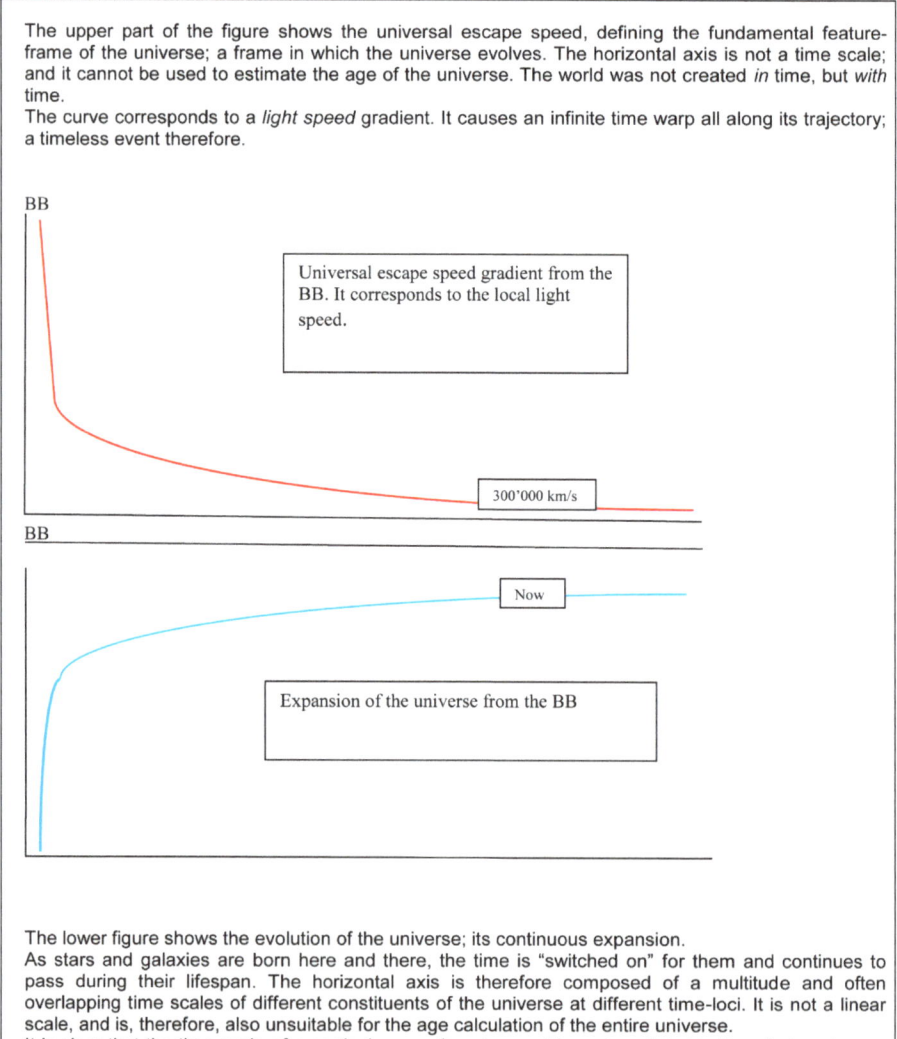

Chapter V

Major Cosmological Events

Soon after the big bang

In the following 2 tables, in order to illustrate the major events since the BB, we have just applied *our scale* of time to the universe's evolution as a whole.

The contemporary hot model of the BB consists of fundamental forces, a variety of different types of elementary particles, and a mathematical description of the behavior of those forces and particles. The forces are gravity, the strong nuclear force, the atomic weak force, and the electromagnetic force.

Figure 5: The first second

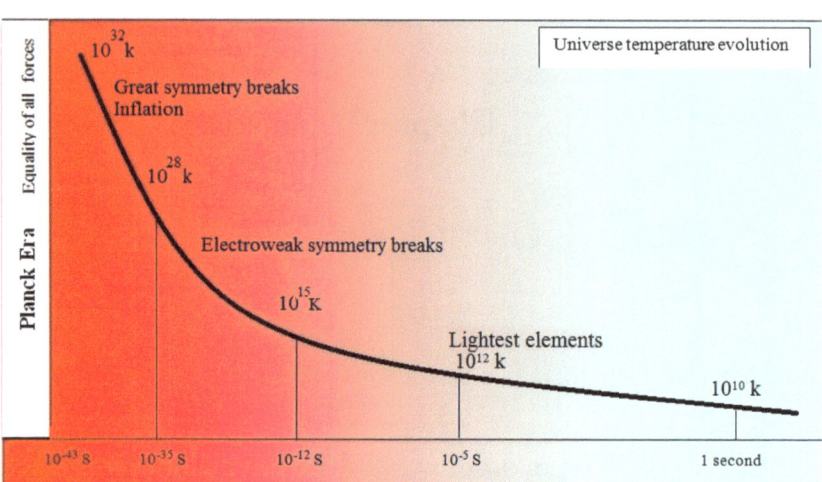

Table of the major events and time periods that make up the big bang model, with *our* time scale extrapolated to the whole universe.

As the universe expands the temperature gradually falls. Although "temperature" is a concept from thermodynamics, it can be related to energy levels because a "black body" at a given temperature radiates photons over a spectrum of energies, but the maximum is at one specific energy level. This energy is taken to be representative of the "typical" photon energy emitted by the black body. A fall in universe temperature is equivalent to a fall in photon energy. Temperature is the measure of particles' agitation. At a very high temperature, particles move so fast as to escape all nuclear or electromagnetic attractions. As a result of frequent collisions, they create particle-antiparticle couples at a rate much higher than their mutual annihilation. The type of particle depends also on the temperature. Within the first second after the BB, the temperature had fallen to some 10 thousand million degrees. This is a thousand times the temperature in the center of the sun. During this period, the universe must have contained mainly photons, electrons, some neutrinos with their antiparticles, and some protons and neutrons. As the temperature fell further with dilatation, the rate of electron-antielectron annihilation increased. More and more photons were produced. Some 100 seconds after the BB, the temperature had fallen to a billion degrees. This is the inside temperature of the hottest stars. This was the period of nucleosynthesis.

At this period, protons and neutrons would not have enough energy to escape nuclear binding forces, and would have started to combine, first producing the nucleus of Deuterium. Further combinations would have given birth to Helium, Lithium, and Beryllium nuclei. In the classic BB model, it is believed that a quarter of all protons and neutrons were converted into Helium nuclei, and a very small portion converted into deuterium and other elements. The remaining neutrinos decayed into protons and formed the hydrogen nuclei, by far the most abundant element.

Then, for a few thousand years, energy dominated matter in the universe. When the temperature reached some 3.5 thousand degrees, the electromagnetic forces could start binding electrons to nuclei and form atoms. At about 3000 degrees, some 350,000 years after the BB, matter-photon interaction ended. With photon decoupling, the universe first became fully transparent, for all practical purposes—light was no longer scattered by matter, and could be emitted as such. This is the origin of *cosmic background radiation*. It also determines the very limit of our observable universe. In fact all of our observed cosmological data are from after this period. Even the Hubble's deepest look into the past cannot go beyond that limit, as light could not be emitted. What happened before cannot be explained by observation, and is, therefore, a matter of scientific theory and debate.

All in the middle

This period corresponds to the phase of continuous expansion, where conditions remain pretty stable, and prosperous to galaxy and star formation.

For whatever it is worth, in this section we have applied *our* local time scale to the rest of universe evolution.

Birth and death of stars and galaxies

The period after photon decoupling is called the period of the "early universe". The deepest cosmological observations cannot go beyond that limit, as light could not move freely before that point. The universe was a very boring place then: there was nothing to see—no stars, no galaxies.

Figure 6: Universe temperature evolution

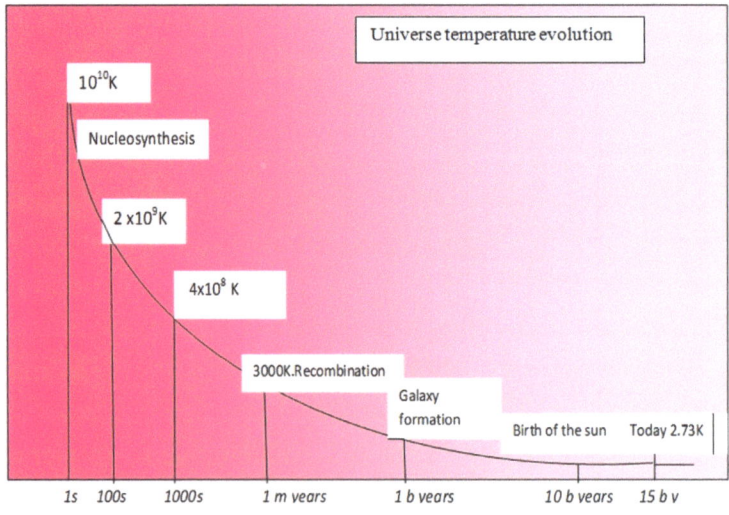

Table of the major events and time periods that make up the big bang model, with our time scale extrapolated to the whole universe.

As the universe continued to dilate and get cooler in a uniform way, some denser regions started to appear. The gravitational attraction in these regions reduced the rate of local expansion, until it stopped it altogether and caused a recontraction, forming the primary nuclei of stars and galaxies. Under the influence of their gravitational attraction and their interaction with the matter in their environments, some of these structures acquired a movement of rotation, which accelerated as the formation became smaller and denser. Rotating galaxies were born.

It can be calculated that the existing hydrogen and helium gas should have been able to condense into the first generation of stars within 100 million years after the big bang, perhaps even earlier. It can also be calculated that the very first generation of stars would have been rather different from typical stars today. They would have been much more massive. But, as with very massive stars today, they would have consumed their fuel in thermonuclear reactions quite rapidly, and so they would have been very bright, with very high surface temperatures and a rather short life. Once they had gobbled up their hydrogen, they would start retracting under their own weight, increasing their temperature and converting their helium into heavier elements like carbon and oxygen. This however, would not liberate enough energy to prevent the crisis of gravitational collapse. What happens at this stage is not very clear; but it seems that the central part of the star undergoes unrestricted retraction into a very dense state—a neutron star or a black hole. The outer surface, containing some heavy elements, may be ejected into space in a gigantic explosion, disseminating matter for the following generation of stars.

Our sun, being a second or third generation star, has some 2% of those heavy elements in its constitution; the rest is hydrogen. Formed from a cloud of interstellar gas some 5 billion years ago, the sun itself absorbed the great majority of the cloud. A minute part, containing the heavy elements, was disseminated by its rotation and formed the solar planetary system, Earth included.

Gravitational collapse and black holes

As far back as 1784, John Michel, a Cambridge professor, introduced the idea that light may be trapped by the gravitational field of a very heavy star. He lectured that there probably exist many such bodies in the cosmos that we cannot see, as their light cannot escape from their surface to reach us, as they are too massive and generate too strong a gravitational field.

Later on, Pierre Laplace, a French scholar working on the same subject, introduced the term "black stars". Much later, Walter Adams, working at the Mount Wilson Observatory, announced that he had obtained a particular spectrum from a star called Sirius B. Sirius A is a bright star in the sky well-known to astronomers, but it has a faint, almost invisible companion (Sirius B), whose existence can only be deduced by the gravitational disturbances it inflicts on Sirius A. The spectrum of the two companions is curiously similar. That means that Sirius B is as hot as Sirius A. The reason Sirius B is so dim is that it must be very much smaller. A small body with an excessive gravitational effect must, therefore, be very dense. At first, the idea of stars squashed into such a small size was not accepted in the scientific community, but over the years it gradually became apparent that extremely compressed stars were not only possible, but inevitable.

Such a dense star could trap light and therefore be invisible.

As we see, the idea of light being trapped by gravity is not new, but its explanation did not come before the theory of general relativity. Einstein's theory predicted that light may be attracted and bent by a sufficiently important force of gravity. These predictions were confirmed later by Eddington and others, who found, during a solar eclipse, a small amount of bending by the gravitational effect of the sun on the beams of light coming from a star situated behind it, exactly as the theory predicted. This was announced and accepted by the scientific community in 1919, during a dramatic joined session of the prestigious Royal Society and Cosmological Society in London. This profoundly amended the old gravitation theory of Newton, which had held for two hundred years previously.

The theory of general relativity relates the strength of the gravitational field at each point in space to the matter and energy that create that gravity. It explains the slowing of the speed of light due to gravity by the warping of time in a curved space-time continuum. A sufficiently small and dense star would then slow or stop light and warp time, depending to its gravitational field. Schwarzchild, in 1916, calculated that if a star of the size of the sun was to be compressed to a radius of 3 km, the time warp in its surface would tend toward infinite and no light could emanate from its surface. The "Schwarzchild radius" is named the "light horizon", an obscure and no-go region from which no beam of light can escape. This state of affaire was contested at first by many scientists including Einstein; but as further scientific data on the nuclear structure and thermonuclear activity gradually became available, the notion of small stars, *white dwarfs,* and *black holes* gained wide acceptance.

In order to understand how a black hole is formed, we have to overview the life cycle of a star.

A star is formed when an amount of interstellar gas, mainly hydrogen, starts to condense under its gravitational attraction. The temperature rises, and atoms collide more frequently and more violently. At first they repulse each other, but, when the gas is hot enough, thermonuclear fusion is set off. Two hydrogen nuclei combine to form a nucleus of helium, as occurs in a hydrogen bomb. The process is called thermonuclear *fusion*, as opposed to *fission*. This is associated with an enormous heat production, which keeps the process going.

As the star starts to glow, its internal pressure rises gradually to compensate its gravitational attraction—its weight. A state of equilibrium is reached, during which the star will use its hydrogen-fuel and produce heat and light. The life span of a star depends on the amount of its hydrogen-fuel at the outset. The bigger the star is at the beginning and the higher its fuel reserve, the shorter its life will be, as it has to use its fuel at a higher rate to compensate for its collapse due to its enormous weight. Our sun has another 5 billion years worth of fuel to shine. When the reserve is used up this equilibrium is lost, and the star begins to cool down and to contract under its own weight.

At first the contraction continues until counterbalanced by the *Pauli exclusion* force: the electron layers do not allow atomic rapprochement beyond a limit. The issue will depend on the initial size of the star. The small variety may reach a state of equilibrium and become what is known as a "white dwarf," with a typical radius of some thousand kilometers and a density of some hundred thousand tons per cubic centimeter. Sirius B seems be an example of a white dwarf star.

Larger stars may overrun the Pauli exclusion force, retract beyond that state, and reach a new equilibrium at the level of the *nuclear exclusion* force. At this stage the electronic envelop is lost and the atom nuclei touch each other. The star's volume has shrunk by a factor of many thousand. At this stage, there are the neutrons and protons in the nucleus of the atom that resist further compression, and a new equilibrium is reached. Those are known as "neutron stars." A typical example would have a radius of some 10 km, with a density of some hundred million tons per cubic centimeter. Neutron stars are small celestial objects with a very strong gravitational field, whose existence was predicted by theoretical physicists, but it took a very long time before they were finally detected.

Finally, largest stars have even a more dramatic evolution. Around 1928, an Indian astronomer, Chandrasekhar, working on the gravitational retraction, came to the conclusion that a white dwarf with a mass of 1.4 times the sun (a limit that carries now his name) would not remain stable, and would collapse further *without an apparent limit*. Nearly all the scientific community, including Eddington, refused the idea of total gravitational collapse, with its associated infinite time warp. Einstein found it unphysical.

It was Oppenheimer (the father of the atomic bomb, and Einstein's boss at Princeton), who, in 1939, described in detail the black hole's "continued gravitational attraction." Using Einstein gravitational equations, he concluded that, given the right circumstances, gravitational collapse will continue indefinitely and the star will plunge on through the *critical radius*, creating an infinite time warp.

Finally, in the 1960s and '70s, in the light of new discoveries and the work of Penrose and Hawking, the notion of "black hole" and total gravitational collapse was definitely entered the scientific vocabulary.

A black hole is formed when a critically sized star has used up its fuel and its thermonuclear activity has ended. As the star begins to contract under its own weight, the g-force becomes so strong that nothing in the world can resist its fierce pull toward the center. In the language of General Relativity, space-time is so much curved that it has reached the rupture threshold and the limit of *elasticity* has been crossed. Any matter in the vicinity of the star—its own material, incompressible atoms, nuclei, and sub-nuclear particles—are all aspirated inward. Even photons, with the inherent speed of light to help them escape, will soon loose their struggle against the force of gravity. They will be trapped in what is known as the "cone of light" corresponding to the *black hole horizon*; a region invisible to the outside world. The zone inside the horizon cannot hold any static matter. Everything in it is swallowed up into its center at the speed of light. The center of a black hole is, therefore, a timeless, empty space—a *singularity*.

Black holes in our concept of universal escape speed

As we have seen in the previous chapters, our concept advocates the continuous expansion of the universe, taking its component members across a diminishing gradient of the speed of light—a gradient determined at the outset by the universal escape speed. This continuous one-directional movement is felt as the passage of time.

At the total gravitational collapse and formation of a black hole, the events appear as if the matter absorbed by the black hole is just "left behind in situ" by the general flow of universal expansion. It has literally hung on the spot. It has not followed the stream of general expansion of the universe and the trend of the diminishing speed gradient. The tremendous force of gravity inside the black hole has accelerated to the limit of the local universal escape speed, all existing accessible material, and created an infinite time warp at the same time. **In our concept of variable speed of light and universal escape speed gradient, anything reaching the local escape speed would be "relocated" to where it belongs. That is a "movement backward in time."**

Or, to phrase it another way, a case of being "left in situ" by the flow of time and general universal expansion is equivalent to a "movement backward in time."

Inside the black hole time has also stopped; and any physical entity must, therefore, cease to exist as it reaches the end of its time. A black hole is a "gateway" to the *end* of time. Anything inside the black hole would, therefore, be "beyond the end of its time."

Although there may be no possibility of any interaction with the inside of a black hole from our space-time, this does not necessarily mean annihilation. Matter, subject to aging, is connected to time. Energy has no time. It looks as if matter reaching the center of a black hole arrives at the end of its time, and is converted into timeless energy.

According to this view, black holes must have been much bigger and less frequent in the early days of the universe, as the gradient speed was higher and it needed a much larger celestial body, with a bigger g-force to undergo a total gravitational collapse in order to form a black hole.

Recently, with the help of Hubble telescope, enormous celestial objects were discovered whose center is occupied by an enormous black hole corresponding to some millions times the size of our sun. These are the famous Quasars, which demonstrate particular cosmological proprieties. These are probably the oldest black holes, formed when the speed of light was higher, needing a more voluminous body to undergo a total collapse. In the outer horizons, where the escape speed is less, bigger and smaller stars will undergo that change at the end of their lives, and the number of smaller black holes must be greater, and be increasing as we go further outward.

Chapter VI

Time

Time, being a major subject and most intimately interconnected with cosmology, is treated separately in this chapter.

One of the biggest and most long-lasting enigmas to confront human thought, the phenomenon of time has always been a matter of philosophical, cultural, and scientific debate. From the dawn of history, the nature of time has proved to be a puzzling and complex phenomenon to humans. In Greek philosophy, the notion of time is often mixed up with the concept of eternity versus transience.

In the old traditional cultures, time was part and parcel of nature. It was recognized intuitively—a vague and subjective recognition. This is evident from the famous answer of St. Augustine, an influential thinker and philosopher of the fourth century, regarding the nature of time: "If no one asks me, I know; but if any person should require me to tell him, I cannot."

The debate over temporality, eternity, and cyclicity of time has troubled human thoughts through ages. Central to the principles of world's major religions, time and its relation to *creation* has led to violent, complex, and heated debates over the generations. Putting God inside the time was not right, as this implied the existence of a period of time before Him: A Godless period. God outside the time induced the notion of incompleteness: A timeless God. What was He doing before creating time?! For how long? Both expressions were blasphemous.

The association of time with dreams, mysticism, and mysterious mythology undoubtedly hindered a proper scientific study of time for many centuries. While topics such as philosophy, mathematics, and geometry had achieved major advances in ancient Greek civilization, the concept of time as an independent existing thing did not emerge until very late. The first notion of time measurement and timekeeping in ancient cultures developed in relation to the rhythmic pattern of the seasons, to day-night alternation, and to the movement of the celestial bodies. It was Galileo who, by comparing the swing of a lamp in a church with his wrist pulse, first discovered the pendulum and its periodicity. He established time as a measurable quantity. Soon the era of clocks and watch-making swept through Europe. But it was not until the work of Newton, in the late seventeenth century, that time occupied a place as a fully independent entity within the laws of nature.

The clock, the emblem of our civilization and of scientific culture today, measures time with increasing precision. Western civilization is entirely built upon precision timekeeping.

The duration of Earth's rotation divided into hours, minutes, and seconds determines the length of our basic units of time. But it seems that the standard measure lacks precision. The Earth's rotation accumulates a delay of almost one second every four to five years, which has to be corrected.

The most accurate clock we now possess is the atomic clock, which is based on the decay beam of the Cesium atom. The central, master clock is located in a well-guarded and supervised laboratory in Bonn, Germany. For a long time it was the "standard clock," the world's central timekeeping unit. Now it is the medium of some two hundred such atomic clocks spread all over the globe, which together represent the standard. The beats are transmitted through the International Bureau of Weights and Measures, near Paris, to all radio signals, and by them to the rest of the time-obsessed world.

A *second* is now defined as 9,192,631,770 beats of the Cesium atom.

The time passes and atomic clocks keep measuring it with utmost precision all years long. One may ask whose time the super clock is measuring, anyway? The Earth's time, the cosmos' time, or the calendar's time?

Mistakes in calendar timekeeping have occurred in the past. The most famous is in relation to the Gregorian calendar. With a decree from Pope Gregory in the sixteenth century, the calendar made a ten-day jump forward. Everybody got ten days older at once! Those ten days, although already lived by everyone without his knowledge, had not been counted, without any bother to anyone.

Newton's fixed time

Aristotle, studying motions, introduced time as a factor. For Aristotle, time was motion. It was perceived only by movements: movements of the clock's handles or those of celestial bodies and earth.

But it was not before the work of Newton, in the seventeenth century, that time came to occupy its crucial position as an independent entity in the laws of nature. Newton brought the measurement of the duration of time to a scientific level equivalent to that of length and weight, and gave it a unit—the second—like the centimeter and gram. He spoke of "absolute" time, which flew on its own, independent of anything else in nature.

Having discovered the laws of motion, he was able to calculate the trajectory of moving objects on earth, the projectiles, as well as the movements of celestial bodies, the moon, and the planets. This represented a gigantic advance in the understanding of the physical world. It seemed that mechanical scientific theories could explain anything in the march of a rational world. A universal, absolute, and completely reliable *time* entered into the scientific language and was accepted, as were the laws of nature and mechanics. Movements of masses under the influence of forces in an absolute space and in an absolute time were the central corpus of the Newtonian edifice. Everything seemed to be explainable by the mechanical laws. The notion of causality and determinism entered the scientific theories.

Fixed and solid Newtonian time lasted for over two hundred years.

Einstein's flexi time

It needed Einstein to tackle the solid Newtonian theory of fixed time and space, and to advocate that time was relative.

Before him, and through the early 1900s, through many uncontested experiments, physics had reached the conclusion that all motion had to be relative. On the other hand, Maxwell's beautiful theory of electrodynamics and other observational data (Michelson's and those of others) indicated a fixed value for the pulse of light, regardless of observer's motion or the source's movement. These two notions were plainly contradictory.

Einstein, by accepting both, proposed a solution that permitted reconciliation. In a brilliant burst of magnificent thought he gave up the universality of *time*, something that had been assumed since the beginning of science. Einstein challenged the notion of a fixed and absolute time: Time is relative; it changes with speed. Speed causes time warp, time dilation. The reason why these variations are not part of our everyday, commonsense experience is that human beings rarely achieve relative speeds approaching even a millionth of the speed of light, and any time dilation is not noticeable in everyday human activity.

Higher speeds, approaching that of light, produce important time warps. Reaching the light speed, time stops altogether!

A clock in the sky….

Is there a cosmic time, unique for everyone and everywhere?

Newton said yes. His opinion held for over two hundred years.

Einstein thoroughly mixed things up with his discovery that there is no universal time, no master clock that monitors the heartbeat of the cosmos. Time is relative. It depends on motion and gravity. But the cosmos is full of both. Time must, then, change from place to place.

This statement is probably not true, as the overall structure of the universe is homogenous on a large scale, and the local variations must cancel each other out. Does this mean that our Earth time corresponds to the cosmic time? And that we can recount the history of the universe contemporaneously with the history of Earth? Nothing is less sure.

On the surface of a typical neutron star, time is warped and slowed down by about 20 percent, relative to Earth. For someone on the surface of such a star, Earth would have 3.5 billion years of age, instead of 5 billion, and the universe 11 billion, instead of 14 billion.

Is the rate of times flow always the same?

We can compare the relative rate of two clocks by placing them side by side, but how can we compare the rate of passage of time measured by different clocks placed at temporal or spatial distances, or in relative movements? Especially when the distances are of astronomical scale?

After the work of Einstein we know that gravitation and relativistic speed influence the rate of the passage of time. The slower aging of the travelling twin at high speed compared to her immobile sister is a classic example in Einstein's theory of relativity. Time has not run at the same rate for them.

As you read these pages now, set aside your book for a moment, close your eyes, and in your mind imagine yourself mounting on the shoulder of a pulse of light. Let yourself be transported for a thirty-minute ride. Look at the landscape. You are everywhere at once, filling up every corner: up, down, right, and left. This is true from the very beginning of your trip. In the landscape before you, you see past, present, and future all at once… In fact it is impossible for you to calculate the duration of your ride. As your trip ends and you return back home the clock in the room has advanced by thirty minutes, while your wristwatch shows the departure time. The time that it has taken the light to transport you from A to B, C, or D has just not *run* for you. At the speed of light, the time warp is infinite. It stopped your watch from ticking. Time had stopped!

How do we know that the super accurate cesium atomic clock that monitors our Earth time will not tick a little faster or slower in a few millions years? Or know what its rate was a few million years ago? We don't mean a particular atomic clock, but *all* of them. How can we be sure that the great clock in the sky has been ticking away evenly from the beginning of time?

Has time changed its rate of passing? If so, what influence has this had on the speed of light?

Our conceptual vision of Time
Past, present, and future

The past is no more, and the future not yet; only the present has a concrete existence!

Is this commonsense illusion or reality?

In our Western society, the classic division of time into past, present, and future seems so fundamental to our experience of life that we take it for granted. To our minds the past, although remembered, has slipped away beyond reach and out of existence; whereas the future, unknown and mysterious, has yet to reveal itself, or even come into being. What is the present, if not a fleeting moment? If that moment didn't move, it would be eternity.

What is the sense of *now*? My now? Does it coincide with *somebody else's* now; somebody who lives elsewhere in the universe? What is the nature of simultaneity?

In the quatri-dimensional space-time continuum of Einstein, time does not flow—it has no direction, not an arrow. The time coordinate is totally similar to the other three spatial coordinates. There is no privileged number or locus, a "now" on the time coordinate, just as there is none on the other coordinates either.

Beginning of time

Even if time is flexible and relative, as Einstein demonstrated, it must have had a beginning and an end. Today, science advocates the BB as the origin of the time. Although the BB is now the orthodox accepted theory, it is nevertheless impossible to provide a convincing account of how the universe can come to exist from nothing as a result of a physical process. The same is true for the origin of time. A complete scientific explanation of how time started is yet to come. Current attempts focus on quantum physics to explain the beginning. *Quantum time* is a complicated and unsatisfactory theory, in which time itself often ends up abolished entirely.

Einstein's revolution got rid of both *absolute space* and *universal time* altogether. Space in the absence of matter is unconceivable in Einstein's physics. But as far as time is concerned, Einstein's theory did not go to the end, giving the impression of an unfinished job. Newton's time was the time of common sense, easy to understand. It was simply there, flowing at a uniform rate in all circumstances. But this simple view was found by physics to be fundamentally flawed. Einstein's theory introduced the notion of an intrinsically flexible time, psychologically deepening its mystery. Time is relative, personal, and subjective, but still obeys physical laws and mathematical regulation. Its rate of flow may be variable in different situations, as in gravitation fields and in relativistic movements.

Our views on time

In our view, time, like space, should also be conceived of *only* in relation with the presence of matter. The relativistic view of Einstein on time should be pushed to its finality, meaning no time in the absence of matter. Time has a meaning only in relation to the constituents of the universe. If the beginning of time is unique—the BB—its end is extremely variable, and is in association with the matter and the constituents of the universe that it represents. All existing matter in the universe has a lifespan. The end of time for an object is when that object ceases its physical existence.

In our thesis and our conceptual vision, division of time into past, present, and future has no physical meaning as far as the universe is concerned. If we accept that at the primary event, the BB, the speed of deployment was that of the universal escape speed all along its trajectory and that this speed corresponded to the local light speed, the time warp would be infinite all the way. Time would simply vanish for the universe itself.

The existence of time and its passing makes sense for the constituents of the universe and that only, during their lifespan. Galaxies, stars, and planets are forming and dying all the time; and these events happen in different *horizons*. During their lifespan, time goes on for them as usual. This is true for all physical objects, big or small, from a galaxy or star to humans and other beings. It is true for atomic or subatomic particles as well. The lifespan for organized biological objects may be on the order of several years to several hundred years, and that of stars and galaxies over millions or billions years. The simpler elements, from Hydrogen to heavier atomic and sub-atomic particles, may have an even longer lifespan, as they may have been around from the beginning, participating in the formation of more complicated structures. Time goes on for and in relation to anything until the end of its existence.

The perception of time by human beings is a matter of the biological activity of our brains. Imagine for an instant that all incoming external sensation ceases. You don't see anything, don't hear anything, and don't feel anything. You have lost the sense of touch, the sense of smell, and the feeling of temperature. Your middle-ear equilibrium system and your multiple internal nerve end organs, situated in different articulations that give you the sensation of position, direction, and weight (gravity), are out of function. More, let us consider that the above dysfunctions have not occurred at the adult age, but have been there from the beginning of your life, so that you have no memory of those sensations. In that situation a brain with no external stimulus will have no sense of existence at all—not for itself nor for anything else.

It is this continuous receiving of physical external signals by our senses that generates the consciousness of "being in the world." The irreversible, unidirectional nature of this continuous memorial coding gives the illusion of the passage of time.

But time passes not only for us, intelligent brain, but also for all other matter: for all the component parts of the universe during their lifespan.

In our view the flow of time is nothing but the continuous general expansion of the universe, which takes with it all of its component parts. This movement is translated into a perceived flow: the passage of time for the physical world.

Three major points to consider here:

> – *The irresistibility* of the flow of time. Expansion of the universe is inexorable, and so, therefore, is the flow of time. Time passes for you wherever you are, and whatever you do. This is true for all animate and inanimate matter. Nothing can resist its flow.
>
> – *The irreversibility* of the passage of time. The general expansion of the universe takes its constitutional elements across a *decreasing* gradient of light speed, defined from the outset: the BB. Light speed is always an upper limiting speed. This forbids any return to yesterday, which had a higher speed limit. There is no way back. Time has an arrow.
>
> – *The variability* of the rate of the flow. The rate of the passage of time is not the same for all constituents of the universe. As we have seen earlier, the fate of the universe itself is determined by the struggle between the opposing forces of attraction and repulsion. It is our view that this struggle will never end, as the intensity of those forces decreases gradually and proportionally.

The universe will, therefore, continue to expand, but at an *ever-diminishing rate* (flat universe at infinity). There are stars and galaxies situated beyond and beneath *our horizon*, our cosmic time-locus. The rate of expansion and, therefore, the rate of the flow of time may be quite different for them. That is why applying our local time scale to the past in order to calculate the age of the universe does not seem to be adequate.

Has time got an end?

Everything in the universe participates in that expansion during its lifespan. Fundamental changes occur, however, when the time of an object is *up*. It reaches the end of its existence. Here we do not mean the end of a biological life for the living, or the end of a physical life for non-living things like planets, stars, atoms, and other dispersed materials in the universe. We also don't mean the decay of atomic structures into sub-atomic elements. All those are transformation. What we really mean is the end of *any* sort of physical existence. Described in the chapter on black holes, the passage through the center of a black hole is associated with the end of time for any matter or structure. These changes are the subject of intense discussions and controversies: the annihilation or disappearance of matter or its conversion into energy…timeless energy.

According to *our* scale of time, our sun will shine for another 5 billion years until it uses up all its thermonuclear material. But, being a rather small star, it will not undergo a total collapse, producing a black hole. It will first blow up into a giant red star, engulfing most of its planetary system, including the Earth, before disappearing with an enormous explosion, a supernova. The dispersed materials of the sun and its planetary system will eventually be taken up by other stars in their formation. Eventually one of those structures may acquire the critical size to undergo a total gravitational collapse at the end of its life; and end up as a black hole, evacuating its material through its center. That would be when the sun's or Earth's material would reach the end of its time.

All matter will eventually reach the end of its time; and matter will be regenerated from energy somewhere else in the universe.

Thus, time has no meaning as far as the universe is concerned, as far as the light speed is concerned, and as far as the energy is concerned. Time is switched on whenever *matter* appears, and reaches its end when matter is no more.

With this view it is clear that *our* "now" may be placed into the future of *some* or into the past of *others* in the universe. Events in the past and future (our past and future) are as real as events in the present. They exist "all at once" across a landscape of time. We have to get away from the notion of things happening in universal sequences. In a real world made of matter and energy we should regard time, like space, as simply there, spread out (past, present, and future), and see those there-and-then events really in existence "out-there."

Consider the lifespan of a tree as an example: it springs out of the earth after it has been planted, it continues to grow reaching its maturity, and lives for many years. Eventually it will reach a period of regression and death.

This is a representation of the *world line* of the tree. Its constituents—leaves, flowers, branches—keep changing all the time. They have each a life-span. At a given time we see only a cross-section of the tree, indicated as *the present*. It is clear that the life of the tree does not consist only of the one-day cross-section of that life, but should integrate all its evolution.

In the same way, in our view, the universe also should be regarded in its entire entity. The current classic model that represents the universe as the surface of a sphere located at some 14 billion years from the BB is restrictive. It shows only a single cross-section of its entirety. The period that fills the inside of the sphere and the period beyond the 14 billion years should also be considered as parts of the universe. The crucial difference with the tree is that the world line of the universe is timeless. The world was created together *with time*, not *in time*.

In a global view and in a timeless world, the universe's evolution, its past, present, and future is laid out all at once.

The past and what has happened in it has not evaporated. It continues to exist. The reason we cannot interact with it is that the past and present each have a *different light speed limit*. As we have discussed earlier, the primary organization that occurred during the BB has left its fundamental footprint in the form of a variable speed limit across the universe. According to the principle of causality we can influence the future by acting now (or when we reach it), as our speed limit is higher than that of our future. But the world in our past had (has) a higher speed limit than our actual one. The inexorable expansion of the universe takes us constantly away, and we could never live in yesterday again; a yesterday in which the limiting light speed was higher. The past world is inaccessible to us any more. To intervene in the past would simply mean crossing the light speed barrier by going back in time: an impossible task.

Where are we?

Light and information from *the inner horizons* reaches us with a decelerated speed; but in much quicker time than it would with our local light speed. Information from the past *is*, therefore, in our possession: we "know" the past! It seems piled up and squeezed into the *present*. We are sitting on it, facing the future. This gives us the impression of being the center of the world and in a privileged position called *the present*.

On the other hand, information from the outer *horizons* reaches us more slowly, and some of it may still be "on the way", so that the future remains generally unknown to us. The past is known, the future not; this seems natural. The present moves onward, linked to the general expansion at a continuously decreasing rate. The continuous expansion of the universe takes all its constituents including us, inexorably across the different *horizons*, whose speed limit, defined at the outset, is gradually decreasing. However small may be the change of speed be in one's life, it is enough to forbid any return to yesterday. The so-called time travel or tunneling into the past, is revealed to be a fiction, as this would mean going faster than the local speed of light, which is the local universal escape speed. Crossing that speed would simply mean leaving the universe, which is clear nonsense. Nothing can leave the universe. Anything material reaching that speed reaches also the end of its time and ceases to exist.

We advocate that this continuous one-way motion is what is felt as the flow of time. Its rate is continuously decreasing.

The final end

The confines of the universe are determined by the end of the gradient of the speed of light, the universal escape speed tending toward zero.

Being timeless, universe has, therefore, no beginning and no end. Instead it has two poles. One of them is the BB, with a tremendous activity, high energy, density, speed, and movement. The opposite pole is where energy, speed, and density all tend toward zero. The weightless photon loses also its energy and stops moving. There is no matter, no movement, no energy, and no time: Absolute nothingness at the limit of the universe: eternal rest and quietness, the "Eternal Promised Heaven".

Bibliography in brief

In order to ensure smooth reading, we have not indicated the source and origin of facts and cited passages directly in the text. We would like to repair this shortcoming by expressing our deepest thanks to those authors and writers whose purposes have been reported here without mentioning their name directly in the text.

Our sources of inspiration in writing this book have been multiples. It would be impossible to mention here the complete list of all articles and references reviewed, especially those found on the web through Wikipedia and other sources. We will mention here the principal books and sources of information and guidance that we have used in writing this book. Our sincere thanks go to their authors.

– *A Brief History of Time*, Stephen Hawking, 1988

– *Sur les épaules des géants*, Stephen Hawking, 2003

– *About Time*, Paul Davies 1995

– *The Mind of God*, Paul Davies 1992

– *Quanta, Relativités I et II*, Albert Einstein, Françoise Balibar, 1989–93

– *Faster Than the Speed of Light*, Joao Magueijo, 2003

– *Si Einstein m'était conté*, Thibault Damour, 2005

– *Sommes-nous seuls dans l'univers?* Jean Heidmann and others, 2000

On websites:

– WMAP Highlights, NASA Explorer 2001–2010, http/map.gsfc.nasa.gov/

– HyperPhysics, Hubble's law and Expanding universe, Google search.

– Science and Reason, the Big Bang,

– Big Bang , article from Wikipedia

– The Hot Big Bang Model, part of Cambridge Cosmology

– The Big Bang, a theory overview from Stephan Hawking's Universe.

– Misconceptions about the B-B, Scientific America, 2005 March.

– The first few Microseconds, Scientific America, 2006 May.

– Expansion of the Universe, Standard B-B Model, University of Helsinki, 2008 February.

– Philosophie et spiritualité, leçon 17, le fleuve du temps.

Author's Profile

I was born and lived in Iran until the age of nineteen. After high school, in keeping with my father's wishes, I came to Switzerland to study medicine. I still remember those first years at the Faculty of Basic Sciences in Neuchâtel, where my preferred subject was physics. I got the highest grades; and the professor asked why I didn't study physics instead of medicine. But I had come to Switzerland to learn to be a medical doctor, and a doctor, I would be.

I successfully finished my studies at the Geneva Medical School, and went to England for specialization in gynecology. Those seven years' of formation in various London and Essex hospitals were very rich and joyful. After the national board exam, I obtained Membership of the Royal College of Obstetricians and Gynaecologists, and I returned to Iran.

At Shiraz University Medical School, a leading and English-speaking institution, I was recruited as an associate professor, and later became the head of the large department of Obstetrics and Gynecology, consisting of some ten faculty members and a program of undergraduate and postgraduate studies in three major teaching hospitals of the region. Those were years of intense activity as, in addition to teaching and the practice of obstetrics and gynecology, we were engaged in two other major issues: that of a change of the curriculum for medical studies, and the foundation of the National Iranian Board of Specialities, with the help and collaboration of the other four major Iranian universities. We had good academic relations with some major universities in US and England. For some of this period, I assumed the function of the vice-dean in charge of education. It was during those years that I was granted the Fellowship of the Royal College of England.

That is to say that during those years I had hardly time to engage my mind to anything else. My cherished physics was very remote those days.

Then came the Iranian Revolution, with great upheaval of norms and values. The medical school was closed and teaching was suspended for three years due to student unrest and group rivalries. Many of the teaching staff left or were fired. Upon reopening the teaching language changed to Persian, and the goals and objectives of the institution were revised.

Some twelve years had passed since my first nomination when I left Shiraz and the academic life to move to Tehran and start a private practice.

By then, the Iraq-Iran war had started; and what was considered at first to be a border conflict, soon spread throughout the whole country. Tehran also had its share of night raids. This war was associated with a long period of insecurity and uncertainty for future.

I was married with two children and, as the war seemed interminable, we decided to move to Switzerland...a painful decision. But that was the absolute calm after the storm.

After the first few years of adjustment, it was in the calm and quietness of Switzerland that I finally found spare time to revive my interest in physics and cosmology. This was first a matter of reading and reviewing bedside books and articles on the subject, and later by taking notes and making a summary of these books and articles. The web and its research tools were of great help.

Finally came the time of my retirement, and the time and patience to put those spread-out thoughts and notes together in the form of this manuscript.

My objective is to share the fascination and wonder I feel for the concepts of the science of the world with other interested people.

www.ingramcontent.com/pod-product-compliance
Lightning Source LLC
Chambersburg PA
CBHW041100180526
45172CB00001B/42